THE STRENGTH IN NUMBERS

The Strength in Numbers

The New Science of Team Science

Barry Bozeman and Jan Youtie

PRINCETON UNIVERSITY PRESS

PRINCETON AND OXFORD

Copyright © 2017 by Princeton University Press

Published by Princeton University Press,
41 William Street, Princeton, New Jersey 08540

In the United Kingdom: Princeton University Press,
6 Oxford Street, Woodstock, Oxfordshire OX20 1TR

press.princeton.edu

Jacket design by Andrea Guinn

ISBN 978-0-691-17406-8

Library of Congress Control Number 2017951861

British Library Cataloging-in-Publication Data is available

This book has been composed in Adobe Text Pro and Gotham

Printed on acid-free paper ∞

Printed in the United States of America

10 9 8 7 6 5 4 3 2 1

CONTENTS

ACKNOWLEDGMENTS

Our book on research collaboration is a research collaboration, one involving not only Barry Bozeman and Jan Youtie, but many others. Much of the work reported here is based on research projects involving a network of valued collaborators and friends who have produced relevant research with us and with one another. We are grateful to, among others, Monica Gaughan, Craig Boardman, Elizabeth Corley, Sooho Lee, Philip Shapira, and Catherine Slade (she is a coauthor of chapter 3). These colleagues are cited extensively in our book, not because they need a boost in their Hirsch Index or even because they are our friends, but because they have produced some of the finest work on research collaboration.

We are grateful to each other as well. It seems a poetic justice that a work on research collaboration effectiveness would arise from one of the most effective collaborations that either of us has had, a collaboration that has spanned many years, generated research grants, and produced both many coauthored publications and a book in which we take some pride, namely this one.

Our research and this book have benefited enormously from government support for our studies of collaboration. Almost all the work reported here was supported by grants provided by the National Science Foundation, most recently from the NSF's Science and Technology Studies program (award number 1026231) and earlier from the Social and Behavioral Sciences Division (award SBR 98-18229). Neither the above-mentioned colleagues nor the NSF is responsible for any mistakes we may have made. Likewise, our opinions are our own.

Several (now former) graduate students have been instrumental to the success of our research. We are particularly grateful to Derrick Anderson (now Arizona State University) who was the "data captain" on the NSF/STS grant. Daniel Fay (now University of Florida) was key in providing literature reports and analysis. Heather Rimes (now Western Carolina University) conducted some of the personal interviews presented here; Youtie, Bozeman, and Gaughan conducted all other interviews.

We are grateful to the staff of Princeton University Press for their professionalism, encouragement and excellent advice. Our editor, Eric Henney, has contributed enormously from beginning to end. We have also benefited from the good work of Brigitte Pelner, Arthur Werneck, Theresa Liu, Stephanie Rojas, and Doreen Perry.

Bozeman owes a special debt to one of his colleagues, Monica Gaughan, not only for her great work as an interviewer, data analyst, research collaborator and frequent coauthor but also for providing sundry, highly valued spousal services. Bozeman and Gaughan are pleased to provide an object lesson, showing that is possible for spouses to collaborate on research and, with considerable effort, remain happily married. Youtie thanks her husband Bill for being her collaborator in life, which has been immeasurably important to her in the production of this book.

Finally, we are grateful to the many academic researchers who took the time to meet with us and be interviewed, or fill out questionnaires, or post information about their experiences on a Website that we developed for that purpose. We are well aware that our respondents had important alternative uses for the time they spent accommodating us. Even though their time with us took away from their work time, every person included in this study treated us with civility, patience, and good will. Their information, good ideas, anecdotes, and, often, their keen wit provided an excellent and enjoyable master class in the social dynamics of science and technology. We are exceedingly grateful. We hope our book does them justice.

THE STRENGTH IN NUMBERS

1

Research Collaboration and Team Science

WITNESSING THE REVOLUTION

Introduction

The scientific myth of the brilliant solitary scientist has long held sway, the image of the scientist emerging reluctantly from his (yes, it is a masculine myth) laboratory to communicate breakthrough results that will push knowledge ahead in great leaps and bounds. However, in recent decades the myth, one that previously held at least a kernel of truth (Lightman 2008), has become more and more difficult to sustain. While there may somewhere be some future Einstein laboring anonymously while developing potentially earthshaking thought experiments, it is becoming increasingly difficult to ignore the fact that almost all contemporary science is *team* science. In today's science, technology, engineering, and mathematics (hereafter STEM) research, more than 90 percent of publications are coauthored (Bozeman and Boardman 2014). Convincing evidence (Wuchty et al. 2007) shows that coauthored research, as compared to single-researcher work, more often leads to high knowledge impacts as well as to commercial uses of research as reflected in patents. Further, the success of collaborative teams attracts more collaborators, thus accelerating the growth of research teams (Parker and Hackett 2012).

Based on years of research on research collaboration and team science, our book aims to increase the probabilities that research teams will succeed in their collaborative efforts. We are certainly not the first students of research collaboration. For decades, others have studied research collaboration, and much can be learned from these earlier studies (for reviews, see Katz and Martin 1997; Beaver 2001; Bozeman and Boardman 2014). So why this book and why now? The succinct answer is that research collaboration and team science are no longer evolving slowly; in the past few years, researchers have seen and actively participated in a research collaboration and team science revolution. The revolution has many aspects, including the growth in the sheer number of collaborators, but also entails a greater mix in the number and disciplinary diversity of collaborators. We are witnessing a new "collaboration cosmopolitanism" (Bozeman and Corley 2004; Ynalvez and Shrum 2011), as researchers from industry collaborate with those in universities, as researchers from one discipline collaborate with those from other disciplines, and as globalization trends and communications technology facilitate increased cross-national collaborations.

While the research collaboration revolution has, in our view, advanced the technological and human resources brought to bear on research projects and problems, it has also created formidable challenges. The revolution presents challenges with crediting and scientific reputation. Historically, processes for assigning credit for research work were reasonably straightforward: a researcher was or was not the author of a scientific paper and was or was not included on a patent. But the traditional norms for recognition break down when the number of authors proliferates. When there are more than a hundred authors listed for a five-page journal article, what does this signify? Related, new ethical problems have begun to emerge. With one or two or a handful of authors, credit attribution presents fewer challenges, but with expanding research teams the likelihood increases that any particular individual contributed literally nothing. The size and diversity of research teams increases the likelihood of conflict. All things being equal, the more persons involved in a team, the more likely that some team members will not play well with others. Research collaboration is no longer about working with friends at the end of the hall or at the other bench in the lab. With the globalization of teams and increased disciplinary, cultural, and gender diversity, we can see that the challenges for research teams differ greatly from the challenges researchers faced pre-revolution.

While almost all researchers are participating in the revolution, some are barely aware of it (most younger researchers take the current research collaboration regime for granted) and others are so busy with their day-to-day work that they have little time, energy, or inclination to spend much time reflecting on the revolution's implications, much less to develop strategies for steering it in the directions they wish. We feel we can help. Our research is on the social and managerial aspects of research teams and the factors affecting research collaboration.

We provide "front lines" reporting on the research collaboration revolution, as well as evidence-based suggestions about how to improve the effectiveness of modern research collaboration. We employ multiple data sources and multiple research methods, including evidence from survey data, data from Web posts, and archival data, but the core evidence presented in our book is from extensive interviews with active, collaborating academic researchers (those interested in detailed information about our data and methods should consult appendix 1). Our book documents and comments on the research collaboration revolution, even as it transpires, and we suggest how research teams confronting a new and radically changed collaboration environment can work more effectively. A necessary first step in coping with revolution is self-conscious awareness—understanding that it is happening, understanding why it is happening, and understanding its components.

Components of the Research Collaboration Revolution

Twentieth-century research collaboration has much in common with twenty-first-century collaboration, many of same advantages, disadvantages, and problems. But there are several elements of contemporary research collaboration that are quite distinct and important enough to characterize a revolution. Revolutionary changes in research collaboration and team science include changes in (1) the sheer number of collaborations and team members per collaboration; (2) commercialization of academic research; (3) gender diversity; (4) multiculturalism and the global conduct of research; (5) increased multidisciplinary (and interdisciplinary) collaboration; (6) contributorship and ethical issues; (7) a self-consciousness about "team science," including new policies and approaches to understanding and managing research collaboration.

THE STRENGTH IN NUMBERS REVOLUTION

Research collaboration[1] is so ubiquitous that it is not possible to understand the dynamics of contemporary STEM research absent some knowledge of collaborative research in teams. Collaboration is nowadays a concomitant of research. However, the idea of "strength in numbers" relates not only to the increased incidence of collaboration but also to the fact that the number of collaborators and coauthors has expanded greatly in many STEM fields. Recently, a paper (Aad et al. 2015) published in the prestigious journal *Physical Review Letters* included 5,154 authors, such a large number of authors that twenty-four pages of a thirty-three-page article were taken up with the listing of authors. We are confident that four-figure author lists will not become the norm. More important is the fact that the number of coauthors per article has increased in every STEM field (Regalado 1995; Abramo and D'Angelo 2015). Even mathematics, the last refuge of the solitary thinker, has witnessed an uptake in coauthoring (Huang 2015).

At first blush, one might well conclude that the increase in the number and incidence of collaborators is an unalloyed blessing. The fact that most studies show that increased collaboration has positive effect on research productivity seems to reinforce this view (e.g., Li et al. 2013; Ductor 2015). However, there are opposing or more nuanced views. For example, Lee and Bozeman (2005) find that different approaches to citation counts lead to different conclusions about the productivity effects of collaboration and coauthoring. With a "normal count" of citations, assigning one citation to each author, the effects of collaboration on citation are quite positive. But with a "fractional count," dividing credit for citations by number of authors, the number of citations accumulated is not greater than for sole authored papers.

From another perspective, it simply makes intuitive sense that research collaboration, despite possible advantages, is at best net positive, not entirely positive in its cost benefit. Research collaborations offer benefits impossible or difficult to obtain, but they also entail costs. The transactions costs (Landry and Amara 1998) in setting up, coordinating, and managing collaborations vary considerably according to a variety of factors, including the number of collaborators, their familiarity with one another, geographic distance, communications media employed, differences in norms, and goals and incentives, among other factors.

In considering the effects of research collaboration on productivity, one may wish to take into account not only counts of discrete knowledge products (e.g., publications, citations, patents) but also more general impacts on research institutions and researchers' careers (Leahey et al. 2015). One of the most obvious problems arising from the increased number of coauthors is the difficulty posed in the evaluation of contributions. When there are, say, eight coauthors, do they all get the same amount of credit? We could say that the first author should receive more credit, but in some fields the authorship is alphabetical, in others it is the corresponding author who is the leading contributor, and in still others it is the last author who has contributed most. The issues surrounding credit and reputation are not merely a matter of scientific ego. In the first place, tenure, promotion, and hiring decisions are all based in part on the reputation and the credit one receives from publishing refereed journal articles. With large number of coauthors, review committees puzzle over contributions. The task is much more difficult when the multiple author problem is exacerbated by multidisciplinary research teams with diverse crediting norms and practices (Lozano 2013; Egghe et al. 2013).

While the complexities of crediting represent an important problem, much more problematic is the phenomenon of "honorary authorship" (Kovacs 2013), instances where people are included as coauthors but who made no contribution to the research beyond possibly serving as a lab director or providing part of the funds for the study or by just being in need of credit to advance one's career. This is not an isolated issue (Greenland and Fontanarosa 2012); one study (Wislar et al. 2011) of honorary authors and "ghost authors" (ones who made a contribution but were not acknowledged as coauthors) showed that a little more than one-fifth of published biomedical journal articles had distortions in the relationship of actual work to authorship crediting, with most of the distortions owing to honorary authors.

The problem of honorary authors has multiple consequences, some beyond the career impacts of the individual. For example, research grants and contract awards are based in large measure on scientific reputation. One might well feel cheated when losing out in the award sweepstakes to a person who has multiple items on the curriculum vita that do not reflect actual work or expertise on a topic. Even more important, persons with large, publications-based reputations are called upon to testify before policy-making bodies and to serve as consultants for industry. When those reputations are inflated by publications in which they had no part, then the expertise claim might be hollow and the advice provided by spurious experts may be inferior (Moffatt 2011; Greenland and Fontanarosa 2012). When the "expertise" is more apparent than real, the consequences are potentially dire (Kempers 2002; Annesley 2011), especially when the topic of concern has to do with public health and safety, such as, for example, the efficacy of new medical treatments or pharmaceuticals.

As we see in later chapters, research managers and professional groups have made some headway with the problem of assessing the individual's contributions to collaborations that include large numbers of authors and even with the thorny problems of honorary and ghost authors. The basic point, however, is that increased numbers of collaborators and coauthors can present problems. There is no likelihood that the strength-in-numbers approach will diminish. Collaboration is a central feature of contemporary research, a revolution in the way research proceeds, and a phenomenon deserving the scrutiny it is receiving from scholars, research team members, and policy makers.

THE ACADEMIC CAPITALISM REVOLUTION

Academic research, our chief focus in this book, continues to reel from another revolution that has deeply affected the very focus of scientific and technical work, the commercialization of research. For decades, US science technology policy insisted that all federally funded research be public domain and, especially, that patenting and drawing individual commercial benefit from such research, that is to say *most* of the research work in academic science, was forbidden. From the late 1980s forward, the policies related to the disposition of intellectual property from federally sponsored research and development (R&D) changed dramatically, in part due to a perceived crisis in national economic competitiveness (Sampat 2006). The Stevenson-Wydler Act of 1980 made technology transfer a mission of federal laboratories and permitted the labs and even lab scientists in some cases to profit commercially from R&D work performed at the lab. In the same year, the Bayh-Dole Act allowed recipients of federal R&D funds, including private contractors, nonprofits, businesses, and—most relevant for our purposes—universities, to file for patents and inventions from their federally sponsored research. While Bayh-Dole was not at its inception viewed as landmark legislation, history shows that it fundamentally altered research universities and their science and technology activities (Grimaldi et al. 2011), having impacts directly on universities' commercialization and intellectual property, as well as wide-range effects on the structure of research institutions, on faculty career motives and performance assessment, on graduate education, and on university-industry relations. Some critics of so-called academic capitalism abhorred these changes as the selling out of universities for commercial goals (e.g., Slaughter and Rhoades 2004), whereas others, especially researchers assessing effects of the "entrepreneurial university," applauded these new activities as more in touch with economic needs, more likely to produce economically relevant education, and as an important ingredient in regional economic growth (for an overview, see Rothaermel et al. 2007).

Later in this book we review the evidence for the impacts of commercially focused university R&D, but at this point suffice it to say that this set of changes qualifies truly as revolutionary and that its impacts on the nature of research collaboration are in some instances seismic in nature. The most obvious change is a vast increase in university-industry research collaborations and, relatedly, the composition of research teams. But commercialization has also in many cases changed or expanded the motivations for collaborating. On the plus side, collaborations often are more fulfilling as a result of new missions and a new mix of actors. On the minus side, legal issues and differences in institutional cultures often pose problems, sometimes thorny and complex difficulties not present in the pre–Bayh-Dole university environment.

THE GENDER DIVERSITY REVOLUTION

The "great man" theory of scientific advance (Boring 1950) no longer has much veracity, though perhaps the "great person" theory works a little better. In 1973, the year the NSF (2014) began collecting and reporting systematic data on the gender mix of academic STEM faculty in the United States, the 118,000 scientists, engineers, and social scientists included 10,700 women, with the largest proportion of women being in the social sciences. By 2010, the year of the most recently available data, the US STEM workforce had grown to 294,800, including 105,200 women—less than 10 percent in 1973, more than 30 percent in 2010. In some fields of biomedical research, parity approaches.

With the diversity revolution, great changes are afoot in the composition and dynamics of collaborative research teams (Leahey 2006; Tartari and Salter 2015; Gaughan and Bozeman 2016). Empirical studies (Fenwick and Neal 2001; Joshi 2014) suggest that gender-balanced teams are more effective in some important respects, but our previous research (e.g., Gaughan and Bozeman 2016) as well as the new results reported here show that gender-based conflict sometimes occurs in gender-mixed research teams and, even more often, gender-related misunderstandings or uncertainty about the impacts of gender on team interactions. A gender-diverse research team often requires a different skill set managerial approaches than single-gender teams.

THE MULTICULTURAL REVOLUTION

Academic science is no longer the preserve of middle-class white Americans. In fact, immigrants have long played an important role in US research and research teams, with post–World War II providing a prime illustration, as Germans and Eastern Europeans, many of them Jews fleeing the Nazi regime, played leadership roles in American physics, including collaborative work in the Manhattan Project. Before that, scientists and inventors from many places in Europe, people such as Vladimir Zworkin (Russia), Ernst Alexanderson (Sweden), James Franck (Germany), and even Alexander Graham Bell (Scotland) immigrated to the United States and changed the history of science and technology. So what is the multicultural STEM revolution? It starts with numbers and geographic origins.

According to the National Science Foundation (2015a), 5.2 million immigrant scientists and engineers resided in the United States in 2013, accounting for fully 18 percent of the US science and engineering workforce. Among these, 63 percent of US immigrant scientists and engineers were naturalized citizens, 22 percent were permanent residents, and 15 percent were temporary visa holders. But Europeans no longer dominate the immigrant STEM workforce. Nowadays, 57 percent of immigrant scientists and engineers were born in Asia, with India alone providing 950,000 of Asia's 2.96 million immigrant contributors to US science and engineering.

As is the case for most revolutionary aspects of contemporary research collaboration, the multicultural dimension provides great advantages in terms of building the scientific and technical capacity of the United States, but it also can present problems in multicultural research teams that must be addressed. One obvious problem is traversing the legal thicket related to visas, especially since US policies differ by country and over time, including in response to national security and diplomacy issues (Wasem 2012). But most research teams, even if they suffer these problems, do not control them. What we find in our research is that collaborations sometimes are held hostage to the very different cultural norms and expectations that some immigrants bring to the team. For example, one common issue is the differences in the status of women and attitudes about women that one finds in the United States versus some other countries, particularly countries in the Middle East and Asia. But sometimes attitudes also differ with respect to such issues as expressing opinions openly, crediting, and even views about intellectual property. The immigrants often bear the brunt of problems, including feelings of isolation and being "second class" members of research teams (e.g., Le and Gardner 2010).

Thus, even though immigrant scientists are on balance a great boon for the United States (Lowell 2010; Kerr 2013; Peri et al. 2015), it is nonetheless possible to encounter cultural differences playing a role in unfavorable collaboration outcomes. These issues have been little studied, but we provide some evidence here that research teams encounter such problems and develop strategies to cope with them.

Usually the term *multiculturalism* embraces not only immigrants who bring the cultures found in other nations but also US citizens who are part of minority cultures. The impact of US minorities on research collaboration or research in general has not yet had revolutionary effects, chiefly because the percentage of non-Asian minorities in US academic science careers continues to be modest. According to the NSF's (2015b) most recent data on the topic, US universities produced 51,008 doctoral graduates in STEM fields (including social sciences) in 2012, up from 40,033 in 2002. In 2002 the percentage of underrepresented minorities (African Americans, Latinos, Native Americans) increased only modestly during that period, from 8.2 percent to 8.5 percent. When we consider the facts that minorities are somewhat less likely to take faculty jobs and that a significant percentage of minorities take jobs in less research-intensive, minority-serving universities, then we see that the vast majority of research teams do not include minorities (except, of course, for Asians and Asian Americans).[2]

Not only can multiculturalism be observed in US science, but also apparent is the rise of international collaboration in science. Wagner and Leydesdorff (2005) show that the percentage of all documents in the Science Citation Index with coauthors from two or more countries nearly doubled from 1990 to 2000. More recently, Science and Engineering Indicators 2016 reported that internationally coauthored publications grew from 13.2 percent to 19.2 percent of all coauthored publications from 2000 to 2013. Although this growth occurred across all science and engineering fields, it was particularly high in geosciences and astronomy. Several explanations have been given for this increase, including the growth of large-scale science around facilities such as particle accelerators or large biomedical efforts related to the human genome, but this explanation is not totally accurate, because international collaboration can be seen in small-scale science, including papers with two or three authors as well. Other arguments include geographic proximity, national research-building policies, prestige seeking, or reduction in the cost of international collaboration through improvement in information and communication technologies. But whatever the region, this international conduct of research has the potential to raise research collaboration issues.

THE MULTIDISCIPLINARY REVOLUTION

Multidisciplinary research[3] is certainly not new to STEM, and neither are the problems flowing from multidisciplinary teams (Thurow et al. 1999; O'Connor et al. 2003). The revolutionary aspect comes from the vast increase in multidisciplinary collaboration (Rylance 2015). Decades ago, working with "strangers" from other disciplines was sufficiently rare that it seemed almost exotic. Nowadays multidisciplinary collaboration is routine, so much so that some researchers have never been on research teams that are *not* multidisciplinary. Porter and Rafols (2009) investigated changes in interdisciplinary in six research fields between 1975 and 2005 and report a 50 percent growth in the number of disciplines whose journals are cited in articles as well as a steady, though modest, increase in the interdisciplinary acquisition and diffusion of knowledge.

A number of factors contribute to the rise of multidisciplinary research, some of them internal to science (e.g., discovery paths) and others related to active encouragement (National Academies 2005) or policy initiatives such as the emergence of large-scale interdisciplinary research centers (Hackett and Rhoten 2009; Boardman and Gray 2010). Nowadays nearly 30 percent of science and engineering faculty at research-intensive universities are affiliated with interdisciplinary research centers (Bozeman and Boardman 2013).

It is easy enough to see the possible benefits of multidisciplinary research and research teams, including the inclusion of different perspectives, skills, knowledge, and even cognitive styles. However, disciplinary diversity poses problems, chief among these an unwillingness to give sufficient respect or trust to scholars from very different backgrounds (Gardner 2013; Ledford 2015). To be sure, with some care multidisciplinary research teams can be managed effectively so as to limit team problems and enhance research productivity (O'Connor et al. 2003; König et al. 2013) but the increase in multidisciplinary research and collaboration inevitably increases the complexity of research teams.

CONTRIBUTORSHIP AND ETHICAL ISSUES IN RESEARCH COLLABORATION

Ethical problems in science are nothing new. While we tend to think of notorious Tuskegee experiments on humans without their consent as an origination point for ethical concerns in modern science, the case was neither the beginning (Lederer and Davis 1995) nor the end (Mastroianni and Kahn 2002) of problems with human experimentation. Similarly, scientific fraud has of late been much in the news (Hein et al. 2012; Steen et al. 2013), but scientific fraud is probably as old as science itself (Gross 2016). However, in the case of ethical issues pertaining to contributorship, there may be something new and, indeed, revolutionary. In the first place, the fact that collaboration and the number of collaborators have increased to such a degree means that ethical issues of crediting come more sharply into focus. In cases of past decades, when papers were single authored or perhaps had only two or three authors, the need for vigilance about those who contributed nothing was surely not so important. In the second place, we shall see in other chapters of this book that contributorship and crediting issues are contingent in nature.

We shall show in this book that various disciplines and fields differ from one another with respect to what is perceived as a legitimate coauthorship; the same coauthoring norms that are standard and commonplace in one discipline may be viewed as unethical in another. The increase in multidisciplinary teams exacerbates disagreement about crediting norms and ethics. Thus, at least some of the issues related to contributorship are new, largely unsorted, and, in our view, revolutionary.

THE TEAM SCIENCE REVOLUTION: REFLEXIVE LEARNING ABOUT RESEARCH COLLABORATION

Public-policy makers and research managers are well aware of the crucial importance of collaboration in the production and application of research, and they seek to enhance the benefits of collaboration directly through public policies developed explicitly to facilitate collaboration (Boardman and Ponomariov 2014). Recently, changes have occurred with policies implemented at the individual project level. For example, the NSF now requires as part of its annual reports documentation and description of research collaborations flowing from their funded projects. On a broader scale, both the NSF and the National Institutes of Health (NIH) have been quite active in promoting collaboration, both by requiring collaborative teams and by setting up research centers with collaborative missions (Roessner et al. 1998; Zerhouni 2003; Boardman and Corley 2008). Policy makers in the United States have also gotten into the act, especially by developing university and university-industry research centers to promote collaboration (Feller 1997; Plosila 2004).

In sum, it is fair to say that funding agencies and policy makers are very much in touch with research collaboration and the role of teams in contemporary science. One of the manifestations of this awareness is the rise of "the science of team science." NIH policy makers and grantees are at the center of this relatively new initiative, one aimed at developing collaborators' reflexive knowledge of research team dynamics and thereby improving the quality and productively of collaborations. This self-consciousness about collaborative teams can itself be viewed as one of the aspects of the research collaboration revolution.

While it is impossible to identify the exact origins of the formal study of research collaboration, it is easier to identify a specific date for the emergence of a very closely related field, the "science of team science." The term *science of team science* is a recent invention, having been coined in 2006 (or, at least, then coming into common usage) as the guiding name for a conference hosted by the National Cancer Institute entitled "The Science of Team Science: Assessing the Value of Transdisciplinary Research." This ad hoc conference led ultimately to an annual research conference and, in 2008, an influential publication (Stokols et al. 2008) in *American Journal of Preventative Medicine.* The emergence of this new nomenclature served to bring attention not only to the new concerns of the science of team science but, at least to some extent, the existing research on research collaboration. However, this development has come at the expense of some degree of conceptual entanglement.

We define research collaboration as "the social process of bringing together human capital and institutions in the production of knowledge." This is not remarkably different from definitions pertaining to team science. Thus, a Website at a recent science of team science conference (http://www.scienceofteamscience.org/scits-a-team-science-resources, accessed May 29, 2016) tells us that research in the science of team science field aims at "understanding and enhancing the processes and outcomes of collaborative, team-based research" and "understanding and managing circumstances that facilitate or hinder the effectiveness of collaborative science, and evaluating outcomes of collaborative science." This definition, then, seems to suggest only a modest difference between the study of research collaboration and the newer science of team science. Arguably, the most important distinction, one that certainly should not be minimized, is the attraction of the vast US health and biomedical research community to the study and application of research collaboration. Even if the distinction between research collaboration studies and the science of team science is one occasioned by old wine in new bottles, new bottles can be very important, especially when marketing products, including knowledge products. A remaining problem, however, is that work on collaborative teams has become somewhat balkanized. At present, health and medical researchers produce the preponderance of work published to date on the science of team science, whereas very closely related work on research collaboration continues to be produced by social scientists (not directly affiliated with health or medical communities), especially researchers in economics, sociology, and public policy.

This division of labor is not, of course, necessarily a problem and characterizes much of the US research and science policy landscape. For better or worse, the "wall" established long ago between the NIH and the NSF (Kraemer 2006; Stetten 1984), the former being the source of almost all funds for biomedical research and the latter being the source for almost all other academic STEM research, affects a great many domains and effectuates all manner of divisions of labor in research, its funding, its policies, and its management. The specific division of labor pertaining to our work, the one between research collaboration studies and the new literature on the science of team science, proves not particularly problematic. To a large extent the respective literatures are dealing with identical phenomena, including exactly the topics we consider in this book. For this reason, among others, we feel that the work presented here is not only relevant to each field of study, it is integral to each of these closely related fields. One of our goals for the book is to bridge this largely artificial divide and to ensure that two closely related fields that have sometimes ignored[4] one another receive some encouragement to cease doing so. This task is made easier by the fact that the research collaboration and the team science literatures have a common "friend." Both literatures are very much influenced by the core (i.e., not directly about science, engineering, or medical research issues) work in organization behavior and group dynamics. There is mutual recognition that many of the lessons in general theories of team collaboration or in contexts far removed from science and engineering are relevant. While there are some aspects of scientific teams that are much different from, say, product development teams or sports teams, or financial teams, there are many other aspects that are to a large extent independent of the particular focus and composition of the team.[5]

While we feel the differences between research on team science and on research collaborations are quite modest, it is worth identifying the few differences that stem from something more than the failure to do a more comprehensive literature search. First, in the case of the science of team science, there is a greater emphasis on what the NIH has long referred to as "translational research," usually meaning the movement of basic and precommercial research findings into new practices, patents, products, technical advances, and treatments (Treise et al. 2016). In some respects this is not much different from historical concerns with the relationship of research collaboration to knowledge and technology transfer or commercial application (see Bozeman 2000; Bozeman et al. 2015).

Another difference between the research collaboration literature and the science of team science literature is that the former is somewhat more expansive. Both literatures are concerned with something more than the immediate dynamics of research teams; both are concerned with context and environmental factors affective collaborative teams. But the research collaboration literature tends to give more attention to institutional actors and large-scale policy influences and, relatedly, sometimes focuses on organizational and institutional levels of analyses (e.g., such as giving greater emphasis to university industry relations). The study techniques also vary a bit, with research collaboration studies using the same qualitative, case study, and survey approaches as team science but also having much greater use for bibliometrics and scientometrics. By contrast, team science tends to make more use of field experiments.

Despite identifiable differences, our basic point is that research collaboration studies and science of team science studies are much more alike than different. Biomedical researchers, just like researchers in physics, chemistry, and engineering, work in teams, face management problems, have concerns with crediting and reputation, and, unfortunately, sometimes face problems in research ethics, personality conflicts, and intellectual property disputes. Most differences are ones of degree, not kind. In this book we use both literatures and the approaches and methods employed by each. In most instances we make no sharp distinctions between the two. Each approach and each literature speaks to issues and problems of collaborative research teams, and each has important lessons about effectiveness. We see our book as a contribution to the literature and practice of research collaboration and, equally, to team science.

In general, the "science of team science" is a most welcome addition to the study of research processes and outcomes, especially because it signifies not only scholarly curiosity (the genesis of much of the work on research collaboration) but also the commitment of policy makers, research managers, and researchers to develop and use of systematic knowledge of collaborative research teams in order to make them better. In this case, self-consciousness is itself revolutionary.

Surviving and Thriving in the Revolution:
Consultative Collaboration Management

The key question posed in this book is: How can one cope with the complexities introduced by the revolution in research collaboration and team science? We have multiple answers to this question, but an important one is related to strategies for managing research teams. In the concluding chapter of the book we present a typology of research management approaches, a typology based on our firsthand observations as well as the data we present here. The approaches include Tyrannical Collaboration Management (fortunately not very common); Directive Collaboration Management (very common); Pseudo Consultative Collaboration Management (with a veneer of democracy painted over a structure of hierarchical direction); Assumptive Collaboration Management (where team members simply assume that all are in agreement about important issues); and Consultative Collaboration Management (not very common but highly effective).

In reviewing these approaches, we discuss their chief attributes and their strengths and weaknesses. However, we come down solidly in favor of one approach, what we term Consultative Collaboration Management. The basic idea of Consultative Collaboration Management is that all team members are consulted at key points in the life of the collaboration (formation, goal setting, task assignment, crediting, disposition and dissemination of intellectual property) so as to identify their respective preferences and values and to decide upon specific actions in pursuit of those preferences and values. We identify more specific elements of the approach and discuss each in detail, elements pertaining to communication structures, assessing team members' contributions, effective means of disagreeing, and identification of diverse objectives, motives, and values. There is nothing magical with this approach, but neither is it in widespread use. However, when it is used, it seems to us to be used to beneficial effect.

Much of the evidence we present in this book is a prelude to our argument for Consultative Collaboration Management. We note here, in our first chapter, that a great many problems in research collaboration are easily avoided. Most problems in collaborative research teams occur not because of malevolence or incompetence but because collaborators assume that other team members share their views and their objectives. However, our evidence shows that most collaborative teams rarely have complete consensus and, equally important, that some members of teams are, for a variety of reasons we document here, unlikely to speak their minds and verbalize their thoughts about the collaboration, especially when their ideas do not accord with other team members.

There is nothing magical about the Consultative Collaboration Management approach. Many of the problems we observe in research collaboration are "textbook" problems of group dynamics (e.g., false consensus, failure to separate judgments of the person and judgment of the idea, a tendency to equate quality in one realm, such as scientific ability, with another, particularly managerial or human relations ability. Solution does not require sophisticated or counterintuitive insights. But the evidence we present here shows that a great many collaborative research teams are essentially on managerial "autopilot" and often to the detriment of the team.

Research Collaboration Effectiveness: Asking the "Simple" Questions

We are concerned with each of the "simple" questions below, but we need to unpack each one. Our questions:

1. What is a "good" or "effective" research collaboration?
2. What are the determinants of research collaboration effectiveness?
3. What can be done to enhance research collaboration effectiveness?

The main reason that these questions are more complicated than they might seem is that different people wish to achieve different things from research collaboration. The multiple objectives for research collaboration mean that identifying determinants of effectiveness and developing approaches to enhancing effectiveness can be a bit tricky. Likewise, and in part because of different objectives for collaborations, researchers do not necessarily have the same concept of the ideal collaborator.

Much of our book is based on interviews, and the ideas and experiences of research collaborators inform all our work. This is a good time to introduce our first bit of researcher evidence, because it impinges directly on what one looks for in a "good collaborator."

THE CASE OF THE ROCK STAR COLLABORATOR[6]

Adam is a guy I worked with at the [gives name of a government laboratory]. He is considered the father of [gives specialty area]. My most cited paper is a collaboration with him. Working with him is like traveling with a rock star. Everyone would visit the lab over the summer. People would come to visit and ask him how he does the calculations. Does he write his own code or buy code? They are surprised to learn that Adam doesn't do those calculations. He has never done anything on his own but he is really good at coming up with the ideas—he is an idea guy—and finding someone else to do that. I was working on some measurements with a colleague and talking with someone outside Adam's office. Adam overheard and he said "these measurements are incredible; we have to write a paper on it". Adam did not really write the paper—I did. But he put his name first. That was suboptimal. But because his name is first, that research is getting a lot of attention. Our papers predated the one that is getting the most research attention, but it came out in *Science* two years after our paper and his paper references his friends. The point is that even if Adam doesn't do anything, a paper he is on will get published just because of his name. A proposal will get awarded just because of his name. So I will include him. In these collaborations—the interpersonal thing is the most important.

So, dear reader, is this a good collaboration, and is Adam really an excellent collaborator? Or is this a bad collaboration or perhaps even an unethical collaboration? Answers depend on who is making the assessment and the criteria being used. Interestingly, the above quote was elicited when we asked, "Can you tell us about the best collaboration experience of your career?"

Thus, we see some of the complications involved in assessing effectiveness. Is a collaboration good because it leads to an excellent outcome, regardless of the behaviors of the collaborators? Is a good collaboration one in which working relations are smooth? Is collaboration effectiveness dependent upon which collaborator is doing the assessing?

With regard to our first question, determining "good" collaborations, one definition of a good collaboration is "a research collaboration that meets its primary objectives, typically objectives involving production of new knowledge or technology." But are all collaborations resulting in a published scientific article best viewed as good collaborations? That seems a low bar. What if the collaboration consumes massive amounts of resources and the scientific results, even if published, are pedestrian, offering little if any significant advance of knowledge? Surely that would not be a good collaboration. Or what if the product of the research collaboration is scientifically first-rate but the process of the collaboration is dreadful? For example, what if one person unfairly seizes all the credit, or if someone of high status or in authority insists on being a coauthor though doing no work on the publication, or what if the collaboration is so rancorous that graduate students involved in the research rethink their commitments to research careers?

Note the passive, impersonal language above: that a "collaboration is a good one if it meets its primary objectives." Subtle but important: this definition treats the collaboration as the unit of analysis, not the individual collaborators. This is not unusual. We often think of research collaborations as entities unto themselves, and in certain ways they are indeed distinct entities, just as organizations, though they are made up of interacting individuals, can be viewed as distinct entities. Even if we think of organizations as distinct entities, most of us have no trouble at all understanding that organizations are social constructs based on the behaviors of specific individuals with specific ties and relationships. In organizations, these diverse individuals sometimes clash and sometimes harmonize, sometimes have conflicting goals and sometimes have converging goals, and some members of the organization contribute productively to the organization's goal, others contribute less. In this respect, most of what we know from studying organizations is directly relevant to the study of the research collaboration entity, the major exception being that the collaborating group may be less formal and is almost always more fluid and changeable.

Just as people in organizations have diverse personal agendas, the same is true of research collaborations. Thus, what is a "good collaboration" for one individual can in some instances be a very bad one for another. How does this happen? It may happen in many ways. For example, let us say that a well-known and powerful principal investigator (PI) and team leader manages a research collaboration team of eight people, all contributing significantly to the creative aspects of the research, and the research is published in a leading journal and is widely viewed as a "breakthrough" article. But let us also say that due to a whim or autocratic decision of the PI, the two neediest participants, say a postdoc and an untenured professor, do not get authorship credit. In such a case, then at least two collaborators may not agree that this is a "good collaboration," even if one genuinely believes that the collaboration has been remarkably successful.

Just one more example should suffice for now. Let us say that an industrial firm has provided substantial funds to a research team, with the express purpose of developing intellectual property that the firm can use. If the research team wishes to publish, that is fine too, but the firm encourages the team to hold off for a few months on publication. The research team takes the money, produces excellent science, but science that has no near-term application at all, then publishes the findings as soon as possible and declares the project both completed and successful. One party to the collaboration (if the sponsor can be deemed a collaborator) will likely not consider this collaboration a success.

These examples show just some of the reasons why it is not always easy to answer the question of what a good research collaboration is. "Good" has multiple meanings, and thus determinants of effectiveness are contingent. What works well for one collaborator may not work at all for another. Fortunately, from previous studies of collaboration (see Bozeman and Boardman 2014 for an overview and critique of the research collaboration literature), we know some of the typical and typically diverse objectives of research collaboration as well as some of the effectiveness contingencies, including such factors as disciplinary composition of the research team, size and geographic distribution of the collaboration, and scientific focus of the collaboration. The importance of such contextual factors also makes it difficult when responding to the final question about approaches to enhancing effectiveness. But having more and diverse evidence helps.

Before delving into the complexities of our data, both qualitative and quantitative, it is useful, we think, to give a summary profile of the questionnaire data on collaboration. This "collaboration arithmetic" is presented below.

Research Collaboration Arithmetic: An Introduction to the Questionnaire Data

As mentioned, our book relies extensively on interviews with researchers, but we also employ questionnaire data based on responses from more than 600 academic researchers working in 108 US universities. The questionnaire data are used throughout the book but provide the primary basis for a chapter focused on decision-making processes in collaboration. Here we present only basic statistics; analysis comes later.

Here we ask, "What are the key characteristics of faculty research collaborations?" We get at this two different ways. In one part of the questionnaire, respondents were asked about the collaboration experiences in the "most recent coauthored research publication." This approach has the advantage of providing a valuable cognitive anchor for the responses—that is, the respondents are thinking about a particular set of experiences around a particular research collaboration that is likely of quite recent vintage. We did not ask them to specify the article; for our purposes it was good enough, though not ideal, that they had in mind one specific research collaboration.[7]

We think this approach proves useful. In the first place, it is not difficult for most researchers to remember the details of their most recent collaboration. Another advantage is that it gives us some insight into the distribution of practices and experiences. If one assumes that examining the most recent publication experience of hundreds of respondents at any particular slice of time represents accurately the diversity of collaboration experiences, then we have a good snapshot. For example, there seems no reason why "the most recent collaborations" should under- or overrepresent collaboration effectiveness for the whole group, nor is there any reason why, in aggregate, the "most recent collaborations" will vary much in the average numbers of coauthors to a paper.

In addition to our interest in this "most recent collaboration," we were also interested in career-long experiences. In a second part of the questionnaire we asked about such factors as career-based percentages of collaborations (versus single-authored work) and possible experience with a variety of negative collaboration experiences. It is perhaps obvious why it is useful to have data looking back at an entire career, but one particular advantage is that some of the negative experiences of interest we knew to be uncommon, the sort of thing that might happen once or twice in an entire career. Since we also had time-based information such as age, rank, and number of years since doctoral degree, having career-based information permitted us to make inferences about the diverse experiences of, say, full professors and untenured assistant professors.

Finally, the full-career questions were likely less threatening than ones about "your most recent collaboration," since it is possible that some respondents were worried about anonymity in the question about one publication in a way they would not be with questions based on their whole careers.

More information about particular methods employed in our survey is provided in other papers (e.g., Youtie and Bozeman 2014), as well as in appendix 2. However, some brief specifications are useful here, at least a sufficient amount as to make clear the nature of the study. Before conducting our Web survey, we needed to determine exactly whom to target for our study. We developed a sampling frame of science and technology fields using the NSF's disciplinary categories in its Survey of Earned Doctorates. Due to previous project decisions, we had decided to exclude the Health Sciences category and Medical Schools. The research collaboration environment in Medical Schools is sufficiently different to render our study too complex. We added one social science discipline to our STEM disciplines—economics—since we thought it would be useful to provide a STEM–social sciences contrast in our study. The resulting sampling frame included, in addition to economics, thirteen disciplines in biology, chemistry, computer science, mathematics, and engineering. The sampling frame called for one male and one female faculty member from each randomly selected department at a given university, because qualitative interviews suggested that gender would be a significant factor; in the event that no female faculty members were affiliated with the department, two male researchers were selected. We sampled specifically for women because a random sample in some fields (e.g., computer science, mathematics) would yield few, if any.

Our sample is from research faculty in STEM disciplines and economics, but only in the Carnegie Doctoral/Research Universities—Very High Research Activity category. We sampled only high research productivity universities, not only because faculty in these universities are more research active and have more research collaborations, but also because most research innovations and breakthroughs and highly cited work comes from those working in these universities. To a considerable extent the Carnegie research extensive universities comprise a different world, one in which research is at least an equal partner with teaching and sometimes, for good or ill, it overshadows the educational mission of universities.

The Web survey, which was concluded in January 2013, yielded 641 nonmedical academic researchers in STEM disciplines in 108 US universities across the United States, for a 36 percent response rate. Respondents were very similar to the population in terms of gender, rank, and departmental discipline.[8]

Who Are the Survey Respondents?

Our survey respondents cover the broad swath of science and engineering researchers. By field, 28 percent are in biology or earth and atmospheric science. Another 27 percent are in chemistry or physics. Engineering comprises 21 percent of respondents, while math and computer science account for 16 percent of respondents. Economics (our social science benchmark) makes up 8 percent of survey participants. The majority of our data come from full professors, accurately portraying the aging of the American professorate (Weinberg and Scott 2013; Finkelstein and Altbach 2014). Of the respondents, 58 percent are full professors, 20 percent are associate professors, and 17 percent are assistant professors, with the remainder being primarily postdocs. Various academic generations are also represented. Of respondents who were PhD recipients, 22 percent received their PhDs in the 1958–1980 time period, 24 percent in the 1981–1990 period, 30 percent from 1991 to 2000, and 24 percent from 2001 to 2012. Most of the respondents in our database are Caucasians (79 percent of the sample), while Asians account for 12 percent, Hispanic 5 percent, and African American/Black 1 percent, with the remainder representing another race (0.8 percent) or preferring not to say 3 percent. The distribution of race is, by and large, consistent with the population for academic researchers in Carnegie Extensive universities and STEM fields.

Now we move from "how" to "what" questions. First we consider the extent of collaboration, then differences between male and female researchers in percentage of collaborative research, the extent of working with students in collaboration, and number of authors on the most recent publication.

What Is the Extent of Collaboration?

Looking at collaborative work during the career, we see that, as expected, most studies are collaborative. As is the case generally, the researchers participating in our survey collaborate a great deal in their research (see fig. 1.1 below). Only about 8 percent of all collaborations during

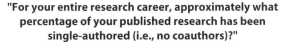

"For your entire research career, approximately what percentage of your published research has been single-authored (i.e., no coauthors)?"

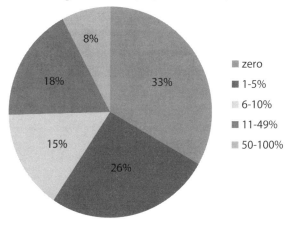

FIGURE 1.1. Career coauthoring patterns.
Source: Research Collaboration Survey. (N = 522 weighted respondents)

the career are single authored. More than one-third of the respondents had *no* single-authored published papers during their entire career. These tended to be younger respondents, especially assistant professors. Another 26 percent said only 1 to 5 percent of their career publications were single authored. Single-authored publications accounting for half or more of an author's career works is very uncommon. Fewer than 8 percent of survey respondents had single-author publications comprising half or more of their career publications, and some of these responses were from very junior respondents with few publications. Single-author publications are much more common in our lone social sciences field, economics. Thus, the notion of academic researchers working by themselves is uncommon (as we expected from previous studies and firsthand experience).

Some of the factors that underlie these results reflect the characteristics of the respondents themselves. Interestingly, female academics were more likely than males to have no single-authored publications (see fig. 1.2). More than 40 percent of female academics had no single-authored publications, compared to 30 percent of male academics. Whether this collaborative nature of female researchers is a result of their positions in the academic hierarchy, their endemic nature, or both is up for debate. It is also the case that female academics are on balance younger, and younger respondents have developed in an environment where collaboration is just a matter of course. But for whatever reason, it is clear that gender differences do exist with respect to collaboration on research papers.

As one might expect, collaboration patterns are somewhat different according to both rank and field. By rank, full professors are much less likely to report no single-authored publications compared to associate and assistant professors. Only 20 percent of full professors have no single-authored publications, while associate professors had twice that percentage (40 percent) and assistant professors had three times that percentage (60 percent). These differences probably do not tell us much about the tendency over time to collaborate but rather about the time required to develop collaborators. Relatedly, many tenured professors collaborate with students and postdocs, which is much less common among untenured professors, who, by and large, give most of their attention to developing a research portfolio that will be sufficient for tenure.

By field, the most striking distinctions concern respondents in the chemistry and physics fields. Nearly half (48 percent) of chemists and physicists participating in our survey have no single-authored publications. These

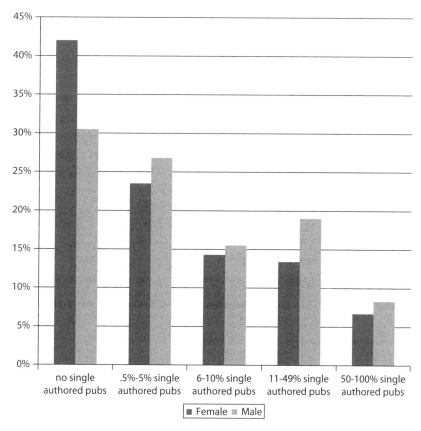

FIGURE 1.2. Male vs. female Collaboration Patterns. Percentage of entire career publications that have been single-authored: males v. females.
Source: Research Collaboration Survey. (N = 522 weighted respondents)

fields, particularly physics, tend to have more involvement in large-scale science. By contrast, only about one-quarter of respondents in biology/earth and atmospheric science (26 percent) and in mathematics and computer science (23 percent) have no single-authored publications. Engineering stands between these two STEM poles at 38 percent. Our social science comparator—economics—shows only 10 percent of respondents with no single-authored publications. Economics tends to by its nature require fewer collaborators (no equipment or large scientific resources) and seems to give greater rewards to the single-authored publication (Ginther and Kahn 2004). Fully 40 percent of the economists who participated in the survey had 50 to 100 percent of their career publications as single-authored publications.

Building the Future: Who Collaborates with Graduate Students?

We asked survey participants, "What percentage of your coauthored papers have included students as coauthors?" Among those who answered this question (519 respondents; a few had no coauthored papers at all), only 6 percent did not have students included as coauthors. The top three most common responses were:

16 percent said that <u>all</u> of their papers included students as coauthors
12 percent said that <u>half</u> of their papers included students as coauthors
9 percent said that <u>90 percent</u> of their papers included students as coauthors

The mean percentage of papers including students as coauthors was 60 percent (standard deviation of 34 percent), and the median was 67 percent. Males and females are nearly equally likely to coauthor with students. Mean student coauthorship percentages are 60 percent for males versus 56 percent for females. Likewise, there isn't much difference in mean student coauthorship rates by rank: 60 percent for full professors, 57 percent for associate professors, and 60 percent for assistant professors. The main area of difference is by field. Chemistry and physics as well as engineering had the highest mean rates of coauthored papers including students, but they also tend to have larger numbers of collaborators on any single paper. The mean percentage of coauthored papers including students was 75 percent for chemistry and physics and the same percentage (75 percent) for those in the engineering fields (taken together). Mean student coauthorship rates are below 50 percent for respondents in biology and in mathematics and computer science, with the former at 49 percent and the latter at 46 percent. Economics again has the lowest mean percentage of coauthored papers, including students at 25 percent.

Numbers of Coauthors on "Most Recent Publication"

To complement our career-wide research collaboration information data, we also have micro-level information about collaboration on a recent coauthored research publication. Thus we switch to a set of questions about the most recent coauthored research publication. By way of background, the most recent coauthored paper had the breakdown of respondents by number of authors according to fig. 1.3.

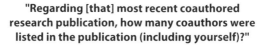

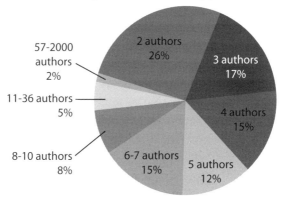

FIGURE 1.3. Number of coauthors.
Source: Research Collaboration Survey. (N = 635 weighted respondents)

In other words, while two-authored publications represent the modal response—with 26 percent publications being two authored—they by no means comprise the majority of coauthorship configurations. Moreover, a few of the papers have "hyperauthors" (Cronin 2001) involving more than fifty authors. The papers examined here tend to be recent ones, as one might expect from a query about "most recent published paper," and they tended to be rather recent. Of these, 71 percent were accepted for publication in 2012 and 16 percent in 2011.

Gender preferences are observed in our survey responses. Male respondents were more likely to work with males in their most recent publication and females with females. Among males, more than half of the coauthors of male respondents were males only. In fact, only 6 percent of male survey respondents did not work with another male coauthor. For females, only 11 percent of the coauthors of female respondents were female only, but only 5 percent of female respondents did not work with another female coauthor.

How cosmopolitan are researchers in terms of their choices of collaborators? Respondents had a tendency in their most recent collaborative paper to work with at least one coauthor at their home university. Consistent with previous findings from our collaboration studies (Bozeman and Corley 2004), most respondents (77 percent) worked with at least one coauthor from the home university of the respondent. But collaborators from other universities are also common, with 80 percent of all respondents having at least one coauthor from another university. Consistent with other studies (Lin and Bozeman 2006; Boardman and Ponomariov 2009), a minority of researchers (18 percent) has even one coauthor affiliated with a private firm.

In the most recent coauthored paper, 26 percent of survey participants were lead authors. First authorship was more likely in papers with only two coauthors (the respondent was the first author in 42 percent of these papers) or only three coauthors (the respondent was the first author in 36 percent of these papers). Otherwise, the respondent was the first author in fewer than 20 percent of the remaining authorship configurations. To some extent, we can expect that the number of authors relate closely to the likelihood of an individual being first author in any particular publication.

In making sense of the lead author phenomenon, it is useful for the reader to understand something we found as part of this work: that "lead author" has different meanings in different fields. In some research fields and disciplines, being the first author is vitally important, in others it is most desirable to be the *last* author, and in still others what is important is being listed as corresponding author. In only one discipline, economics, is it common to have author order alphabetic. We come back to these issues, which can prove surprisingly thorny and contentious, in our later chapter on "contributorship" and coauthor decision-making.

The Rest of the Book

By this point, the reader should have a good idea of the basic objectives, approach, style and overall tenor of the book as well as some initial insight into the data we employ. While each of the chapters provides a distinctive contribution, they share a common goal: the desire to shed light on research collaboration effectiveness in general, and either by implication or empirically based "lessons for practice," to suggest steps researchers can take to improve their own collaboration experiences. One of the primary contributions of the book is to provide conceptual tools for thinking about research collaboration. We first pursue this goal in chapter 2, introducing a simple typology of research collaboration outcomes, emphasizing that most research collaborations are positive but that the ones that are not are often quite damaging. We use our interview data to illustrate the range of research collaboration outcomes. For the pessimists among our readers, we show that some research collaborations can be viewed as Nightmare Collaborations, where everything seems to go wrong and with dire consequences, including exploitation by the powerful of the less powerful or less experienced, unethical behavior, personal attacks, and such. These types of collaborations oftentimes lead to such outcomes as lifelong professional recrimination, law suits, besmirched reputations, and people abandoning research altogether. The optimists among our readers will be gratified to find just how uncommon are Nightmare Collaborations and will be buoyed to learn that the vast majority of research collaborations are quite successful, especially in terms of the perceptions of participants.

We expect this book will be of interest to two quite different constituencies, the large group of researchers interested in improving their research collaborations and the smaller group of researchers, chiefly social scientists, who study either research collaboration or related social and economic aspects of science, technology, and research. Chapter 3 provides a succinct review of literature on research collaboration, one that includes the most recent findings about research collaboration and the science of team science. Many of the readers in group two, those conducting research on collaboration and related topics, will find their work discussed in chapter 3 but may also benefit from the organization and synthesis of the literature. The larger group, those interested more in managing and improving their own collaboration experiences than in the nuances of research and theory about collaboration, should be able to tease some relevance from the various results reported in chapter 3. Both readership groups may wish to consult our appendix 2, in which we provide a companion literature–based propositional inventory. This inventory is both more extensive and, at the same time, less so. It is more encompassing because it provides even more research findings about collaboration, but there is little commentary or assessment, and in that sense it is more limited. We hope that the appendix provides a readily accessible tool for science and technology studies researchers while at the same time it furnishes a quick reference for researchers seeking results related to specific collaboration issues they may be encountering.

In chapters 4 and 5, closely related to one another in both purpose and style, we begin to get serious about in-depth presentation and analysis of the data we developed for this study, beginning in chapter 4 with an analytical framework we feel is one of the more important contributions of the book and then applying the framework in chapter 5 to the data. Chapter 4 introduces our Aggregate Model of Research Collaboration Effectiveness and chapter 5 applies it, chiefly with reference to the interviews developed for this book. The Aggregate Model of Research Collaboration Effectivness identifies a number of determinants of research collaboration effectiveness, largely based on the research literature reviewed in chapter 3. The model includes collaboration management processes as a key factor, one that receives relatively little attention in the research literature but one that is featured here as one of the major instrumental approaches to improving collaboration effectiveness. Why focus especially on collaboration management? Since our book is focused on the individual researcher, with the assumption that the researcher will be interested in learning about and possibly improving his or her own research outcomes, we take special care to emphasize those factors over which the researcher has some control. Thus, while such factors as government and university policies may have strong effects on research collaboration experiences, most researchers do not have much control over these factors. By contrast, researchers often play a major role in their collaboration choices and the management of collaborations.

While chapter 5 makes extensive use of our interview data, chapter 6 focuses on our survey data and examines the motives, activities, and decision processes of the respondents' reporting about their research collaboration experiences. The two key parts of the chapter are responses to, first, questions about respondents' most recent collaboration and, second, about the collaborations they have experienced throughout their career. The data show that the vast majority of researchers experience some negative collaboration outcomes, but only a minority of specific collaborations entail bad outcomes (such as colleagues who did not produce work promised, extensive delays, crediting issues, exploitation, or gender conflict).

Our concluding chapter, chapter 7, is in some respects the most important, because it is here that we draw some lessons from our data and research about how specifically to improve research collaboration teams. In this chapter, we distill lessons from the Aggregate Model but then present an observation-based typology of research collaboration management, including the Consultative Collaboration Management approach that we feel has potential to increase the likelihood of effective collaborations.

Consultative Collaboration Management, generally our preferred approach, is not the most common approach to collaboration management. That distinction rests with Directive Collaboration Management, whereby one of the collaborators is, essentially, the person in charge. The directive manager may or may not consult other team members on important decisions but, in any case, provides the last word on decisions about such factors as collaborator recruitment, specialization, and crediting. Why is

Directive Collaboration Management so common? The approach reflects hierarchies one finds in science and engineering research and practice. Often the directive collaboration manager is the laboratory director, the principal investigator, or the dissertation or postdoc supervisor and easily gravitates to the notion of being in charge, generally with the assumption and often the reality that others acquiesce.

A less common collaboration management approach is the one we refer to as Tyrannical Collaboration Management, an approach that in some ways resembles Directive Collaboration but can be thought of as its pathological counterpart, characterized by desire to dominate and little or no respect for others' opinions. Tyrannical Collaboration Management is often associated with the Nightmare Collaborations we discuss in chapter 2 and elsewhere in the book.

There is no need here to summarize all the collaboration management approaches identified in chapter 7, but it is worth underscoring that any of the various approaches to research collaboration management, with the possible exception of Tyrannical Collaboration Management, can under the right circumstances be effective. While we feel that Consultative Collaboration Management generally offers the best hope of enhancing effectiveness, the real key to effectiveness is fitting the management approach to the needs and the resources of the specific collaboration. Thus, for example, a team of collaborators who are experienced, who have collaborated with one another frequently, and who have easy access to one another's time and attention will have very different needs than will a team that includes people not well acquainted with one another, who have diverse perspectives and backgrounds, and who have different statuses. In short, one size does not fit all, and as we review the findings throughout this book, that is a good point to bear in mind.

2

Routine and Not-So-Routine

CLASSIFYING RESEARCH COLLABORATION OUTCOMES

"A happy marriage is a long conversation. . . ."
—ANDRE MAUROIS

Introduction

In chapter 1 we considered some not-so-simple "simple" questions about what is a good collaboration. We have already seen some of the complexities in determining just what is a good research collaboration outcome and what is not. As a first step to bringing some clarity before considering all the complexities of research collaboration and the many determinants of effectiveness, we begin with a modest analytical grouping of research collaboration outcomes: Routinely Good, Routinely Bad, Dream Collaborations, and Nightmare Collaborations. We use this terminology throughout the book. Given all the complexity in assessing research collaboration processes and outcomes, it is useful to at some point rely on this simple, intuitive distinction. One should bear in mind the point that most research collaborations succeed, as evidenced by the fact that researchers keep collaborating throughout their careers, many times with the same collaborative teams. One means of emphasizing the fact that most research collaborations succeed: our "Tolstoy Principle."

The Tolstoy Principle of Research Collaboration

We give disproportionate attention to the minority of problematic research collaborations, dissecting them in an effort to determine what typically goes wrong when research collaborations fail to meet researchers' expectations. The problems research teams face are diverse and may include, among other possibilities, problems in the management and social dynamics of the collaboration, occasional battle of the sexes, exploitation of the weak by the strong, and sometimes, more rarely, clear-cut ethical violations. We give more attention to the negative, not because we are inveterate pessimists but because we feel that much can be learned from failures. If we understand failures, then it is sometimes possible to address them and increase the likelihood that future collaborations will meet with more success. Thus, even if bad collaborations are no more than 5 to 10 percent of all research collaborations, the bad ones are well worth studying. Not only can we learn from failures, but the worst failures have dire consequences, including instilling distrust, perpetuating bad behavior under the belief that it is normal behavior, and sometimes even causing some disillusioned researchers to abandon promising careers.

Given that the book examines so many of the problems occurring in research collaboration, it is a good idea to remind ourselves often that bad collaborations are very much the exception. Our whimsically named Tolstoy Principle provides a good mnemonic for helping us remember that most research collaborations succeed and most research teams are "happy ones."

Tolstoy Principle of Research Collaboration: All happy collaborations are alike; each unhappy collaboration is unhappy in its own way.

We think the happy family cliché has at least as much veracity for the metaphorical families of research teams as for actual kinship and marital families.[1] However, even as a generalization the Tolstoy Principle requires some elaboration, the most important of which is this: happy collaboration *families* are happy in the same way, but family *members* may have different preferences among aspects of happiness. This may seem contradictory, but actually it is not. Let us explain the difference between happy research teams and collaboration outcomes and happy team members.

Much of this book is based on researchers' responses to questions about "good" or "effective" research collaborations. We have confidence in the validity of these responses and in the shared meaning among researchers.

Not a single person asked us, "What do you mean by 'good' or 'bad'?" Not a single person struggled to answer questions about good or bad collaboration experiences. This does *not* mean they have exactly the same specific preferences but rather the same general preferences, with differences being ones of degree. Considering all the evidence gathered for this book— interview responses, Web posts, survey questionnaire responses—it is clear that when researchers talk about good research collaborations, the collaboration invariably includes the following components:

1. *Knowledge outcome*: There is a positive scientific or technical outcome (e.g., a quality publication or an important patent or increased resources).
2. *Harmony and trust*: Working relationships are generally harmonious, either based on trust or generating trust; relationships may entail some disagreement but disagreements focused on enhancing work quality, not personal acrimony.
3. *Scientific and technical human capital*: Researchers improve their careers and capacity to produce work; the collaboration enhances their reputations, career prospects, and ability to succeed in their profession.
4. *No shirkers*: All members of the research team make significant, positive contributions to the research collaboration.
5. *No exploitation*: All team members are dealt with fairly in terms of communications, expectations, and crediting.

We find no disagreement on any of these points. True, some would add other elements to their concept of a good collaboration: for example, some list "having fun," whereas for others it is all business, listing factors such as "lively intellectual exchange" or "complementary skills." But not one respondent disagrees about the five points identified above (even if they do sometimes disagree about the extent to which they have been achieved in any particular collaboration).

Agreement about the core components of a good collaboration does not mean that all collaborators have exactly the same intensity of value for each component. Some are willing to trade a little disruption and conflict for greater scientific impact, others seem to prize team harmony over nearly all else, and still others evaluate effectiveness in terms of particularistic aspects of scientific and technical human capital development—such as, for example, developing students' talents or giving a postdoc the skills needed for a good permanent job. In short, and Tolstoy notwithstanding, when we discuss "good collaboration" we cannot stray too[2] far from the accompanying questions "why" and "for whom." Thus, our *modified* Tolstoy Principle: *All happy collaborations are alike, but collaborators are not all made happy in exactly the same way.*

The unhappy family uniqueness in our Tolstoy principle works best with Nightmare Collaborations, the *really* bad collaborations, ones that are not merely unpleasant but have long-lasting negative effects, that damage reputations, produce psychological scars, and sometimes involve costly and time-consuming lawsuits. If they are resolved at all, the unhappiest research collaborations require the expenditure of much time and energy and sometimes money, and they have long-term effects on future collaboration choices, including sometimes the choice of avoiding collaboration when possible.

Most very bad research collaborations generate interesting, awful stories and, in the worst of cases, embarrassing news media headlines. But most of the stories are interesting both because they are rare and because they differ, in terms of outcomes, not only from routine collaborations but also from one another (as we shall see below). Good, effective collaborations make collaborating team members happy. Usually there is sameness to the stories of good collaborations, and accounts can even be a bit boring, as researchers lavishly praise their best collaborators as kind, understanding, hard-working, productive, and dependable. In effective collaborations, research team members enjoy working with one another, they produce a quality knowledge product (e.g., article, algorithm, patent, grant proposal, conference paper, technology), no one is exploited, every participant benefits in some way important to him or her, and—an excellent diagnostic of effectiveness—collaborators are eager to work with one another in the future. Fortunately, such outcomes are commonplace. If effectiveness were not the rule, were research collaborations scattershot and unpredictable experiences resulting often in frustration and unhappiness, then research collaborations would doubtless be less common and the scientific results from collaborations thereby diminished.

We provide evidence throughout the book that collaborators have different goals for collaborations, sometimes compatible goals and sometimes not, and these are realized to different degrees for different collaborators. But for now, let us work with the fact that most respondents have a general concept of effectiveness and that there is at least some shared meaning to their ideas about effectiveness. Thus, the typology below gives us a handy way to characterize collaborations, a way that does not permit nuance but is easily communicated.

A Typology: Routinely Good, Routinely Bad, Dream, and Nightmare Research Collaborations

Figure 2.1 distinguishes among four types of research collaborations: Nightmare, Routinely Bad, Routinely Good, and Dream. The figure gives a crude pictorial representation of their relative frequency in the distribution of all collaboration outcomes (we provide more precise statistical evidence later in the book).

Routinely Good Collaborations are by far the most common, comprising the vast majority of research collaborations. In such collaborations, research teams and individual collaborators achieve scientific, career, and personal goals with little rancor, no bad outcomes, and in general the experience is a pleasant one that all parties may wish to replicate with future collaborations. In Dream Collaborations the experience is even more memorable and positive. For example, the participants may find they tremendously enjoy working with one another, the collaboration exhibits excellent complements of skills, participants have compatible work styles, they are responsible and come through on agreed- upon tasks. In such collaborations, participants almost certainly wish to continue working with one another, at least if scientific goals and circumstance permit. Researchers speak quite fondly of such memorable collaborations. They do not happen often, but when they do the experience energizes all those involved.

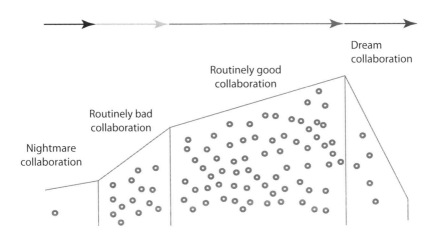

FIGURE 2.1. Hypothetical distribution of research collaboration outcomes.

In Routinely Bad Collaborations, at least some significant parts of the experience are vexing, but not to such a degree that there are major long-term consequences. For example, one colleague might be a credit hog, or a collaborator may do a slapdash job in his or her work responsibilities, or one may be consistently late with promised work. Routinely Bad Collaborations may induce greater concern about future collaborations among the same individuals but do not necessarily preclude working together in the future.

Nightmare Collaborations go horribly wrong, but they can unravel in a remarkable variety of ways. Nightmare Collaborations have deep impacts that negatively affect careers, either objectively or in terms of job satisfaction, sometimes even resulting in changes in careers or significant changes in research agenda or collaboration choices. Nightmare Collaborations in many cases involve lawsuits or allegations of unethical behavior. Nightmare Collaborations produce not only animosity but also enemies. The good news is that *most researchers never experience a single Nightmare Collaboration.* Indeed, as we get more deeply into our interview and Web-based evidence, we shall see that there are only a few collaborations that seem clearly to be in the Nightmare category, but a few are on the borderline.

Table 2.1 summarizes the typology of research collaboration outcomes and provides some elaboration of the concepts as well as a few indicators in each case. Using this typology, we found that it was easy to classify more than 90 percent of the collaborations—at least from the perspective of the respondent—as falling into one of the discrete categories.

TABLE 2.1. Typology of Research Collaboration Outcomes

Outcome category	Distribution density	Suggested indicators	Magnitude of impacts
Dream	Uncommon, but experienced by most at some point(s)	-Highly successful from scientific and social process standpoints (e.g., breakthrough publications; excellent working relations and greatly enhanced social capital) -Make or change careers and views of collaboration, enhanced reputation -Very high level of satisfaction reported from all parties to collaboration -Great enthusiasm about future collaboration	Major, positive

Outcome category	Distribution density	Suggested indicators	Magnitude of impacts
Routinely Good	The most common, the vast majority of collaborations	-At least moderate success from either scientific or social standpoints (e.g., publications in refereed journals, increments to social capital). -Modest positive impact on career -Moderately positive satisfaction reported by all or most parties to the collaboration -Encouraged to collaborate in the future	Minor, positive
Routinely Bad	Common, experienced occasionally during most careers	-Work proves ineffective due to such as loss of funds, team members' inability to "deliver," poor working relations, or poor management of collaboration -No significant impact on career -Moderate dissatisfaction for some or all parties to collaboration -Careful weighing of any future collaboration opportunity	None or minor negative
Nightmare	Extremely uncommon, many never experience	-External mediators required -Lawsuits -Careers threatened and sometimes ruined -Extremely high degrees of antagonism -Some parties become bitter enemies	Major impacts, extremely negative

Illustrations from Our Data

In this section, as through much of the book, we learn directly from researchers' reports of their experiences in collaboration. The collaboration experiences presented below are diverse, but they represent each of the four categories.

DREAM COLLABORATIONS ILLUSTRATED: TWO MENTOR-CENTERED COLLABORATIONS

Here is an example of a research collaboration that seems to qualify for the Dream category. The respondent was extremely enthusiastic describing this collaboration experience. The case involves a mentor, as so many Dream Collaborations do:

> Early in my career I met a senior chemist (I'm a biologist) that helped launch my interdisciplinary career via collaboration on science, on

proposals, and even by paying for me to attend an important international meeting when I did not have funds to attend as a starting faculty member. His example has shaped my behavior toward many junior students, post-docs, junior colleagues, etc. over the years and I've now helped many junior colleagues in ways that I might not have been wise enough to have done without having seen the role model he set. Many of those people are now senior and I see them doing similar things. Thus, my senior colleague's initial support has been "paid-forward" through multiple generations of faculty.

Among Dream Collaborations, this is not at all an unusual account. In this study but also in previous survey-based research we find that a significant number of senior researchers are quite serious about the mentoring role in research collaboration, and, indeed, for some it is one of their primary motives for collaboration and one of the primary determinants of the choices of collaborators. An earlier study from this project (Bozeman and Corley 2004) examined researchers' motives for collaboration. While the study found a variety of motives, including different or complimentary skills, similar work habits, and even working with someone with a "big scientific reputation," one of the most common motives was to help or to mentor junior colleagues and students. In their questionnaire-based study with data from 1,041 university-based STEM faculty, the authors found that more than 75 percent of senior faculty respondents listed mentoring and professional assistance as one of the most important of their motives in collaboration. In our view, sincere and committed research mentors provide remarkable resources for individuals and for entire fields, often passing along good and beneficent practices to new generations of collaborators. The pay-it-forward notion expressed in the above illustration is vital, and those committed to it often form the backbone of Dream Collaborations.

Since the mentoring collaboration is such an important and especially productive type of dream collaboration, we consider another instance, this one a response from a mid-career biologist:

This is about an especially positive collaboration. When in graduate school, my thesis advisor began working with researchers outside of our immediate research field. This led to a team of six of us, two full professors, one associate, and three postdoc students, representing three different universities, working on research together. My advisor and I (and another grad student) provided methodological and theoretical expertise, and the other three provided content knowledge expertise (for educational research in the sciences). Four years later, we have one grant, two papers, half a dozen conference talks, a dissertation, and a pending grant proposal, all developed from this collaboration. We are now at six different universities (the students are now assistant professors) and we continue to try to find ways to work together. I think it works well because we are all interested in the research, have a fair amount of energy and excitement about work, we respect each other's expertise, we trust each other to do what we say we'll do, and we genuinely get along with each other.

While mentors often play a prominent role in Dream Collaborations, we shall see in other sections of the book that Dream Collaborations occur at various points of careers, they include relationships not only between students and mentors but also among career peers, and they have diverse outcomes, not just scientific publications and grants but also excellent working relations with industrial partners, for example. Dreams are not nearly so diverse as Nightmare Collaborations, but neither are they confined to a single story thread.

ROUTINELY GOOD COLLABORATION: ONE GOOD PROJECT LEADING TO ANOTHER

We do not spend a great deal of time on Routinely Good Collaborations, chiefly because they are so common, so similar, and, frankly, not so interesting or dramatic. No one tells stories around the campfire about their pretty good collaborations. Still, they are important because they are the grist for scientific and technical progress and for career advancement and, most important, because most collaborations are Routinely Good.

Here is an example of a Routinely Good Collaboration, one based on a theme that could be replicated in close form in nearly every academic unit in nearly every program and university:

> We had completed the measurements [in our research project] and saw some interesting results. We contacted a colleague that we had talked to at many conferences, but never collaborated. He was invited to give a seminar, we discussed the data, and he did the calculation, communicating with the student [on the project] back and forth. Then we realized that to make the comparison to the calculation we needed better characterization. We spoke with a colleague at our university and he directed us to a new faculty member, who greeted us with great enthusiasm and said

why not make it a collaboration. Since I was fully tenured and pressure of "diluted" contribution has been removed, we went for it. It was an excellent experience. We learned so much and had so much fun that we are now starting a completely new project together.

Okay, this is not a riveting story. But that is typical of Routinely Good Collaborations: they tend to be good, small things that are satisfying but not life altering.

ROUTINELY BAD COLLABORATION: THE CASE OF THE FLYBY COLLABORATOR

Routinely Bad Collaborations take many forms, and thus we provide several examples here. Many of the collaboration problems discussed in our book are faculty-student problems that often flow from the student's lack of experience or confidence or from the faculty member's expectations or inability to anticipate actions of inexperienced collaborators. Here is one such case. For reasons soon apparent, we refer to it as the Case of the Flyby Collaborator:

> John, a faculty colleague and I, had decided to collaborate on a paper for a conference. The paper was accepted and we started working on it. It was not difficult to produce a good paper because we had resources from a grant, good data, and the ideas were already developed in the grant proposal. So we quickly put together a good paper. The problem was that due to various work and personal developments neither of us was actually able to go to the conference, which was in Canada, a long way from us, and would kill two or three days we did not have at the time. However, we came up with a solution. We had a senior doctoral student, Robert, who was soon going on the job market and who could benefit from the conference and he could present the paper. We made him a co-author based on the fact that he had developed some very nice tables and figures for us. This would normally earn an acknowledge-ment, not a co-authorship, but he was needy, the graphics and the figures were fantastic, and we would all benefit by his going to the conference and presenting the paper. So, he went to the conference for us. About a month later I happened to see a very good friend who had attended the conference, a good enough friend that she did not mind telling me that lots of people at the conference were very upset that I neither showed up nor told them that I would not be attending. Bad form! Naturally I

thought this was some sort of misunderstanding and that the student made the presentation but neglected to identify his co-authors. I talked to the student with the idea of providing a mild admonishment that he should publicly identify co-authors. He became highly agitated, very upset, and finally told that while he did fly to Canada for the conference, once he got there, he practiced the presentation but then decided that presentation was really not of the quality he would like (I had seen it, it was fine). So he was so embarrassed that his presentation was not as good as others at the conference then he just stayed in his hotel room during the panel. Since his name was not actually on the program, since the switch was made late, John and I managed to collect our colleagues' wrath, while Robert collected a nice trip.

Next we have a very different Routinely Bad collaboration, one focused on a troubling problem in authorship crediting.

ANOTHER ROUTINELY BAD COLLABORATION: THE UNPLANNED, UNWANTED HONORARY AUTHORSHIP

Research collaboration and coauthorship is increasingly beset with crediting problems, including the practice of "honorary authors," ones who have their names on papers but contribute little or nothing. This case is an interesting twist on honorary authorship. While many cases of honorary authorship are best classified as Nightmare Collaborations, this one is a Routinely Bad Collaboration because it entails no dire or lasting consequences:

A few years ago, I was invited to give a research presentation in Taiwan and, while there, I was hosted by a very bright, likeable professor and his primary master's student who was assigned as my overall driver, tour guide and general purpose aide-de-camp, obviously with the intent that I would learn a lot about her, be impressed with her, and support her admission to our department's doctoral program. When I returned home, her admission was easily achieved, she was impressive and no favoritism was required. Shortly after she enrolled in our program I happened to take a job at another university but we stayed in touch. I also stayed in touch with Professor [gives name] since we had many common interests. Now, flash forward three years. I am at my home and receive a notice that I have a package I must sign for and I go to the post office to pick it up. The package is U.S. currency, about $200. It is a large package and also includes what seems to be a journal. I could not make much sense of the journal, all in Chinese, except that I could see in English characters my name in large font in front of one of the articles. I then discover a handwritten note from my now former doctoral student and my Taiwanese colleague Professor [gives name] telling me that they are pleased to have me as a co-author on the paper published in this Chinese journal (I never was able to read the name of the journal). The note said that it is customary for the journal to pay for contributions and my share of the pay is enclosed. Apparently, my contribution to their paper, unknown to me, was the set of tables and figures from the PowerPoint slides I presented during my trip to Taiwan. What should one do in this case? I donated the money to our university's scholarship fund and then simply wrote them an email with a "thanks."

We provide another example of a Routinely Bad Collaboration, one that almost everyone experiences in their career, though not usually to such a degree as this case:

ROUTINELY BAD COLLABORATION: THE DELINQUENT CONTRIBUTOR

As we shall see in the survey data provided in later chapters, no complaint about collaborators is more common than their not finishing promised work or providing it late. The vast majority of Routinely Bad Collaboration reports are owing to these difficulties. However, the case below provides an interesting twist:

When I was a beginning faculty member at [gives university name], I began a collaboration with my friend Roger. Roger was a personal friend from a different department. We had some mutual interests but I had never worked with him. We knew each other well, played on the same softball team, same poker group, a good guy, so why not work together? I hit on an idea for a paper and I thought he would be an excellent collaborator. I already had developed most of the data and, since I was familiar with the data and he was not, I figured I would do most of the work setting up the research design, the hypotheses and the data analysis and Roger, who was not really that comfortable with statistics, would develop a related conceptual model, write the literature review and the paper's introduction. . . . Seemed a clear-cut allocation of duties.

A month passed and I finished all my work and passed it along to Roger. Hearing nothing after another month had passed, not even an acknowledgement that the material had been received, I went to his office with a simple "did you get the stuff." He assured me that he not only received it he was making progress and I would have his contribution soon. Finally, we are less two weeks from the time we have to mail in the paper and, so, I call him and say "look, I know this is a busy time for you" (actually I don't think it was but I gave him the benefit of the doubt) "so I can take over much of this." He responded that he would be very grateful and he says "after all, you are the first author." He asked to cut his duties back to providing an extensive literature review and bibliography.

I finished up everything except his literature review and then sent the package to him, a little more than a week before we have to send off the paper. The next day I am very sick with the flu, not "flu-like" but the real thing. I am no longer thinking about the paper, I'm thinking about keeping my toast down. In two days, though, I am starting to feel better and, so, I call my collaborator and ask about the literature review. "Sure, sure, I will bring it right over." I am now well enough to be sitting up in bed, propped up by pillows. He comes into the room, wearing a surgical mask, and he says "I think you will find this very useful" and puts a big folder on my bed. The folder contains more than 50 photocopied abstracts of journal articles. "Great," I say, "and where is the literature review file?" He says, "oh, this is everything you will need for the literature review."

Before leaving the realm of the Routinely Bad Collaboration for the more dire and consequence-laden Nightmare Collaboration, we provide one final case because it underscores the very important point that "bad" is not always clear cut and, relatedly, that collaborations can be very good in some aspects and very bad in others.

NOT ROUTINELY BAD, NOT ROUTINELY GOOD: THE "BEST WORST COLLABORATOR"

Gender dynamics often play out in collaborations, sometimes but not always with a senior powerful man bullying a junior and less powerful woman. There are many dimensions to gender dynamics, and one special case arises from collaborating spouses. Professors tend to marry other professors, they tend to marry people around them—often in the same university or program—and because they have so much in common and because of the frequent interaction and proximity among other reasons, they often collaborate on research.

Consider the comments of a tenured (but still early career) male physics professor:

> Who is my "best" collaborator? My greatest collaborations have been with [gives name]. She is right there in the office next door. Even though she has a different last name she is my wife. Our collaborations have been very successful. However, when I work with my wife it is torture. We fight all the time. But our arguments seem to make very good papers. It is also probably good for our marriage, because we seem to not have any time to fight about anything else because we fight so much about our research work.

While the gender dynamics of spousal research collaborations present problems not precisely the same as in other collaborations, there are similarities to less special cases. Researchers need not be married to enjoy and suffer all the consequences of pooled labor in close proximity. Many close collaborations, including ones occurring over several years, generate conflict and pain as well as pleasure. When people work closely together, they sometimes develop dynamics not at all unlike marital partners: continual sore points, which are irritated often, but also a willingness to overlook relatively small slights in order to achieve big benefits. In many respects, exchange theory (Emerson 1976) explains as much about research collaboration (Li et al. 2013) as it does about marriage (John et al. 1976; Clark and Mills 2011). That is, individuals in both types of social relationships make decision about their investments of time, resources, and caring based on calculations of what they will receive in return.

NIGHTMARE COLLABORATION: THE STALKER "MENTOR"

As mentioned above, Nightmare Collaborations are so uncommon that with all our questioning, and with our Web post invitation, we managed to uncover only a few. But sometimes they occur, and they are in some cases not difficult to classify. Here is one such unequivocal case:

> During my geophysics post-doc, my direct supervisor decided he was in love with me and began to stalk me. This caused a lot of anxiety and I began to have health problems (shortness of breath, heart palpitations, and stomach problems). When I reported the man to a senior manager, somehow the problem became public. I had to leave the research group, and then was denied access to work I'd done before, and my name was removed from papers resulting from our collaborations.

This is a Nightmare Collaboration not only because of the sexual harassment but also because the respondent's attempt to address the sexual harassment problem resulted in punitive action that had a direct impact on her academic research productivity and reputation. This led to a series of unfortunate institutional responses that led to significantly altering the trajectory of this scientist's career (the passage below is from the same interview):

> I was never able to integrate myself into the research group I was placed in. At that point, I left my PhD field and took a student position in [gives name of new work group] at the same institution. It was a wonderful group and I enjoyed my time there. However, I had to start over again when my husband and I came to our current institution. I did yet another post-doc in [gives new field] and am now a senior researcher [in a nontenure track position]. Things have been somewhat better in this field, but far from perfect. Again, there are few women, and non-professorial research staff (mostly female) tends to get the short end of the deal, irrespective of the scale of their contributions. I suppose I am a little bitter about it, after trying so hard for so many years to gain respect and acceptance. Occasionally I wonder if I should have left research and gone to medical school years ago—at least I would have some job security and a little respect.

The above passage is the key: the chief element in Nightmare Collaborations is that they have long-lasting negative consequences affecting careers and quality of life.

THE BOUNDARIES OF NIGHTMARE COLLABORATION: THE CASE OF THE VANISHING COLLABORATORS

As mentioned, Nightmare Collaborations are quite uncommon, and some of the few we have been able to identify are not so clear cut. Consider the following example, from one of our Web posts, a case related to a significant authorship-crediting problem:

> The worst experience that I had was collaborating with some MDs (I'm a PhD in biomed and I've worked a lot with vets and human doctors). In this case it was a project initiated by the MDs. They had the initial ideas and funding. They came to us for help. My student and I helped them a lot with project design, experimentation, and data analysis. Then we discovered that they had submitted our work to a research conference without putting our names on it. Then they asked us to help them prepare

their presentation. They didn't even think they had done anything wrong! I got a more senior faculty member to bring that up to them. They had clearly just used us and weren't thinking of us as collaborators even though we had done most of the work and they couldn't even interpret the findings without us. They did later correct it and add us to the paper.

Is this a Routinely Bad or a Nightmare Collaboration? Not an easy call. However, the fact that the collaboration involved exploiting a student and the apparent level of disrespect exhibited could well imply that the level of badness meets Nightmare standard. But the basic point: the line between bad and really bad is sometimes a fine line.

SOME IMPLICATIONS OF THE EXAMPLES

Though the implications of the above cases are likely clear from the above commentary, it may be useful for us to summarize them. First, most research collaborations succeed, as one might well expect given the popularity of collaboration. We make the point, with our Tolstoy Principle, that effective collaborations have much in common. However, *even in the case of highly effective collaborations we cannot conclude they are successful in all aspects.* There are many dimensions to research collaboration effectiveness, and there is no necessity, nor is it even likely, that every single dimension of a research collaboration will prove to be a highly positive one.

We noted here that Routinely Bad Collaborations are relatively uncommon and Nightmare Collaborations are so rare than many people never experience even one during their careers. In a later chapter we provide quantitative evidence that this is so.

A point not discussed above, but worth introducing now, is that every field and discipline has good and bad collaboration outcomes and that some outcome types are distinctive in one particular field or discipline while others are common to all disciplines. It is easy to see why this might be the case. A field that relies on large-scale and extremely expensive equipment could have access to that equipment as an important determinant of collaboration effectiveness (Shrum et al. 2007). Similarly, in a field that relies on scarce samples, the management of those samples could be vital (Manolio et al. 2007). Physics or biomedical research teams sometimes have issues because the teams include forty or fifty collaborators or more (Cronin 2001). This is not (at least yet) an issue in collaborations among economists (Bidault and

Hildebrand 2014), in which case a "large" collaboration is one with three people.

Yet despite the distinctiveness of fields and disciplines, there are collaboration issues and problems that are common to all researchers in all sciences, "hard" or "soft," experimental or mathematical, qualitative or quantitative. Thus, for example, economists and physicists have ghost authors, chemists and sociologists have colleagues who promise much but deliver nothing, and computer scientists and anthropologists have personality clashes with colleagues. However, just as there is particularity to research collaboration, there is also universality, as we show below.

THE UNIVERSALITY OF RESEARCH COLLABORATION EXPERIENCES

Surprise (perhaps): two of the illustrative cases above are from the authors' own experiences,[3] but we feel that the experiences relate not at all to our field, discipline or training and they could have happened just as easily to persons in any STEM field. While there are different issues according to field or discipline, collaboration dynamics and outcomes have much that is common that is distinctive, even when we are comparing the social sciences with the physical and natural sciences. Thus, it is not necessary to come up with, say, a theory of research collaboration effectiveness for optoelectronics and a different theory for chromatin gene expression. Discipline and field are important contingency factors, they affect outcomes, but that does not suggest that disciplines' problems and collaboration dynamics are unique. There is no need for entirely separate theories or different approaches to effectiveness.

With the exception of a subset of respondents who are economists, both our interview data and our questionnaire data is based on responses from researchers in the physical and natural sciences, mathematics, and engineering. STEM[4] researchers' collaborations differ from most economists and social scientists in a few important ways. In the first place, STEM researchers collaborate at higher rates and usually with a larger number of collaborators. In some STEM fields it is virtually impossible to make it as a solitary researcher; this is not yet the case in any social sciences fields. Interestingly, with larger numbers, the likelihood of problems increase, not only due to management problems and "transactions costs," but also to the fact that increased numbers of dyadic relationships give increased likelihood of all sorts of social outcomes, including personality clashes and different work styles. Some STEM researchers differ from most social scientists in that they are often heavily dependent on equipment, samples, gene sequences, and other laboratory accouterments, often sharing resources or negotiating resources controlled by others. Vitally important, the intellectual property stakes are often huge in some STEM fields but almost never significant in the social sciences. All of these are important differences.

STEM researchers probably have most in common with social scientists with regard to the human problems experienced in collaboration. Even if social scientists' work differs in important ways from scientists' and engineers' work, they nonetheless experience quite similar conflicts, lapses in collegial behavior, and the intrusion of real life onto research life. Whether collaborations are based on data from synchrotrons, from living tissue, or from living and lively interview respondents, the dynamics of human and group interaction are quite similar in collaborating groups.

RESEARCH ON RESEARCH COLLABORATION

It should be clear enough that our book is not one that is aimed chiefly at researchers. We seek to illuminate research collaboration practices and ultimately to help practicing researchers improve their collaboration. Thus, the book is not deeply embedded in social science theory, and in the interest of ensuring accessibility we have taken pains to avoid giving much attention to the technical aspects of our research. While we make use of the literature on research collaboration and team science, we do not review it extensively. Thus, the next chapter is an exception. While social science researchers are not our primary audience, they are part of our intended audience and in all likelihood are more interested in aspects of literature and theory. For more practical-minded researchers, the next chapter may not be "required reading," though indulgent readers will find at least practical lessons from scholarship on research collaboration and team science.

3

The Literature on Research Collaboration and Team Science

Most readers are more concerned about improving their own research collaborations than acquiring a detailed understanding of the research literature on collaboration and its implications for theory. "Applications readers": you may wish to take a pass on this chapter or to keep it in reserve as a future resource. Of course, many of the findings of the research collaboration and team science literature have direct implications for improving effectiveness, but some of these lessons can be absorbed from reading the other chapters of this book. For "scholar readers," those interested in research and theory of collaboration because it is the object of their own study or because of intellectual curiosity, this chapter's treatment of the literature may have value, especially since it is up to date, at least as of this writing. In appendix 2 we present a propositional inventory table, and this table can serve the application readers' concern with finding specific insights into specific research effectiveness issues, while at the same time helping the scholar reader develop a more comprehensive and detailed knowledge of the literature.

The Focus and Organization of the Review

Fink (2010) suggests that literature reviews explain the need for and significance of the research. Others (Blaxter et al. 2006; Blumberg et al. 2005) suggest that literature reviews not only summarize existing materials in a given field but also include an assessment of current knowledge. Following these guidelines, we provide a literature review of research collaboration focusing on issues examined throughout the book, including:

- What Is Research Collaboration?
- Assessing Research Collaborations with Systematic Thinking
- Norms and Decision-Making Processes in Research Collaborations
- How People Work in Research Groups
- Gender Dynamics and Research Collaborations
- The Institutional and Professional Contexts of Research Collaborations
- Contributorship Issues in Research Collaborations
- Promoting Better Research Collaborations

For those interested more in specific findings than a more holistic review and commentary on the literature, our appendix 2 may prove more useful; it is a propositional table organized by subtopics, to help the reader identify relevant literature concerning research collaboration effectiveness.

For this chapter we draw extensively from our previous work and our work with colleagues, particularly a monograph by Bozeman et al. (2013), "Research Collaboration in Universities and Academic Entrepreneurship: The-State-of-the-Art" in the *Journal of Technology Transfer*, and a recent book by Bozeman and Boardman (2014), *Research Collaboration and Team Science: A State-of the-Art Review and Agenda*. But this chapter represents new work and includes a good deal not in those previous reviews.

What Is Research Collaboration?

It should be clear from the foregoing chapters that we focus on the question "what is research collaboration?" not because we wish to belabor the obvious or because we are obsessed with small differences but because the meanings are sufficiently different in the literature to warrant caution in interpreting and synthesizing findings. Studies on research collaboration often have many different and ambiguous meanings, and in some unfortunate cases otherwise valuable studies neither provide a definition of research collaboration nor make their meaning clear from context.

One major conceptual ambiguity in defining research collaboration is easily addressed by defining the level of analysis. The chief focus for this book is on research collaborations among individuals and not among organizations. We also recognize that it is not always easy to distinguish individual collaborations from organizational collaborations. But after all, when organizations collaborate, it is actually the individuals in the organization who are relating to one another. Organizations are such a part of daily life that it is sometimes easy to forget that organizations are convenient social constructs based on patterns of human behavior and individual relationships. When researchers are asked to identify their collaborators, unless they are explicitly asked about their organizational relationships, they tend to identify individuals as collaborators, not groups or organizations (Youtie and Bozeman 2014). Still, we note that relationships between organizations, as well as relationships of individuals within organizations or social network analysis (Burt 2000; Clark and Mills 2011), often are important features of collaboration and merit some attention even in a book such as ours, which is focused on collaborative teams.

Many researchers, particularly those in research universities, tend to think of collaboration in terms of coauthorship because coauthorship is conveniently measured and, not least, because authorship is the currency through with they are employed, retained, and advanced. Thus, much of the published work about research collaboration focuses on coauthorship. As Katz and Martin (1997) point out in one of the best-known and most comprehensive reviews of research collaboration, the coauthor concept of collaboration has several advantages, including verifiability, stability over time, data availability, and ease of measurement. However, they note that coauthorship is at best a limited indicator of collaboration and there are many other dimensions of collaboration to be considered and evaluated.

We go beyond the Katz and Martin approach and suggest that coauthorship is but one of the many outcomes of social processes encompassed by a research collaboration. In our view, coauthorship is neither necessary nor sufficient for constituting a research collaboration. In chapter 1 we defined research collaboration as "the social processes whereby researchers come together jointly to deploy their human and social capital for the collective production of scientific and technical knowledge," a definition similar to the one used by Shrum and colleagues (2001) in one of the best-known works on research collaboration. By this definition, collaboration need not be focused on publishing articles, and, indeed, collaborations often are more concerned with technology development, or software and patents, and may have no publication objective at any point.

Thus, while research collaborations result in an identifiable knowledge product such as a scientific paper or a patent, by our definition there is no implication that collaboration will necessarily succeed or even that it will be brought to full term. Sometimes when people collaborate, the collaboration fails or it is abandoned before providing a knowledge product. These fallow efforts are nonetheless collaborations. Coming together "jointly to deploy their human and social capital for the collective production of scientific and technical knowledge" does not imply that the goal is achieved. Thus, some researchers (e.g., Shrum et al. 2007) measure collaboration according to shared goals and resources for a specific project, not according to the outputs from the project. There are many other reasons why research collaborations do not bear fruit, especially when academia-industry collaborations with intellectual property considerations are involved (Flipse et al. 2014).

We prefer our definition of collaboration, which is broader than coauthorship, broader even than production-focused definitions, because it recognizes that collaborating researchers sometimes go down blind alleys and that these "failed" collaborations can prove beneficial by, among other reasons, ;ruling out certain paths and thus conserving time and resources; (2) proving pathways to new and potentially more productive research topics; and (3) bringing together researchers who learn about one another, possibly using this knowledge to facilitate future collaborations (Youtie and Bozeman 2014). Since research is inherently unpredictable, a concept of collaboration that accommodates unpredictability offers advantages.

By our definition, research collaborations are about human capital and relationships between people, not coordination of other resources. Obviously, financial resources are vital to the success of many research collaborations, but our definition suggests that one who *only* provides resources is not a research collaborator but a patron. Sometimes patrons become coauthors, and sometimes this coauthorship without human capital contribution can present problems; thus, by our definition patrons are not collaborators (Bozeman et al. 2013). In our view, the patron-as-collaborator notion is one of the chief contributors to increased numbers of honorary authors, a development that most see as highly undesirable (Greenland and Fontanarosa 2012).

Our idea of human capital is an expansive one, though it focuses on scientific and technical human capital (Bozeman et al. 2001; Ponomariov and Boardman 2010), not human capital in general. A person who has knowledge of laboratory equipment may bring that form of human capital to a relationship and, by our definition, be a collaborator. For example, Stokes and Hartley (1989) showed that sometimes a researcher might be listed as a coauthor because he or she has provided material for a study or performed an assay in a laboratory. However, in some cases an individual who makes a major contribution may neither obtain nor desire coauthor credit. For example, a mentor may help shape a vital part of a doctoral student's dissertation, perhaps even providing the core idea, but never expect or receive more than an acknowledgment in a publication (Müller 2014).

These broader notions of collaboration, while intriguing, are not often easy to measure, and focusing on coauthorship alleviates many measurement problems. Thus, many useful studies for defining research collaboration (e.g., Heffner 1981; Vinkler 1993; Melin and Persson 1996; Wagner 2005; Heinze and Bauer 2007; Mattsson et al. 2008; Mayrose and Freilich 2015) begin and end with the coauthored publication. Despite the challenges of measuring productive research within a broader concept of research collaboration, our book and this chapter review and indeed emphasize collaboration concepts and research studies that go beyond coauthorship. Most of these studies (e.g., Melin 2000; Bozeman and Corley 2004; Bozeman and Gaughan 2011; Youtie and Bozeman 2014) rely on researchers to nominate collaborators, explain the nomination, and describe the collaboration relationship. As such they help broaden the definition of research collaboration beyond coauthorship (Jeong et al. 2011). Other studies describe research collaboration as a division of labor among collaborators (Laudel 2001) to define creative contributions to science policy.

Assessing Systemic Factors in Research Collaborations

Our book focuses chiefly on the individual level of analysis, meaning relationships among individual researchers, rather than relationships among and between organizations or the relationship between individuals within institutions and organizations. Though a challenge, some studies work at more than one level of analysis at the same time (Hou et al. 2008).

The literature on research collaboration effectiveness would surely be improved with more studies that simultaneously work at different levels of analysis, integrating knowledge of individual collaborators with knowledge (and data) about their systemic and institutional contexts. Most current research on collaboration tends to either (1) focus intensely on individual researcher issues while ignoring the larger context within which the researcher operates; or (2) focus on collaborating organizations at a level of abstraction sufficiently general as to permit no consideration of the role of individual dynamics that may shape the outcomes of collaborating organizations (Bozeman et al. 2013).

As do others (Vasileiadou 2012), we understand why system thinking, with multiple levels of analysis, is a big problem; the analytical requirements and the data requirements for many levels of analysis generally are prohibitive. However, progress is being made with respect to system thinking about research collaborations and their effectiveness, and the literature is growing with respect to discussions of typologies used to explore the system approach to research collaborations, meaning research by whom and for whom (Hessels 2013).

One conceptual framework focusing on both systemic and individual level factors is the "contingent effectiveness model" we previously developed (Bozeman et al. 2015). In this model, multiple parties to research or members of a system have many goals and many definitions of effectiveness, including those that relate only to public values. Effectiveness can include what is accomplished, the market impact, any impact on economic development, the result for political advantage, contributions to development of scientific human capital, and any opportunity costs. We see the increasing complexity of the system approach to assessing research collaboration effectiveness. This broader effectiveness concept is consistent with the work of Shrum and colleagues (2007), who demonstrate that scientific and technological collaborations are part of a general trend toward more fluid, flexible, and temporary organizational arrangements (Ulnicane 2015) and the need to study long-term research collaborations with system implications (Bozeman et al. 2013; Bozeman and Rogers 2002).

Narrowing our focus to development of scientific and technical human capital, we are seeing growing literature on the social relational aspects of research collaboration, conceptualized by Emerson and colleagues (1976) in terms of social exchange theory. They describe effective teamwork and behavior exchange between team members. The approach used in some of the recent work on the science of team science seems similar in focus, even if not driven by social exchange theory per se (e.g., Gray 2008; Bennett and Gadlin 2012; Lotrecchiano 2013). At the same time, other contributions to the team science literature have led the way in either research on or, more often, conceptualization of collaboration as a systemic and multilevel phenomenon (e.g., Börner et al. 2010; Falk-Krzesinski et al. 2011).

Mitchell and colleagues (2015) are among those who view research collaboration from a more systemic perspective, focusing on concerns about the cost of research collaboration in terms of power struggles between collaborators, time spent on developing collaboration, conflict between collaborators, stress, process issues, suboptimal outcomes in collaborations, and resources required to build and manage collaboration teams. Li and colleagues (2013) describe the importance that researchers place on relationships with colleagues, given the demands for academic achievement. Defining several indicators of social relationships between researchers, they find that systems of researchers and outcomes of research evolve based on social relationships and issues such as trust and distrust between collaborators.

In an interesting approach to system-level thinking, Gray (2011) considers cross-sector research collaborations to better explain the major institutional participants in scientific research systems. He notes in particular the many formal and informal linkages that are productive despite the fact that they evolved from a highly structured set of relationships between industry and university partners controlled by federal funding initiatives.

We similarly note the significant role of government and policy systems in shaping relationships between researchers. Clark (2011), for example, examines the role of the federal government in shaping the relationship between academic scientists and their industry colleagues. He finds that for federally funded projects, increasing collaboration between academics and their industry colleagues leads to less academic-to-academic collaboration. This approach resembles the earlier work of Hagstrom (1965), who warned that academic-to-academic collaboration and more risky long-term projects might be threatened by systems of funding agencies and governmental priorities.

These systemic features of collaboration are examined in more detail in the first appendix to this book, the propositional inventory of research collaboration and team science studies.

Norms and Decision-Making in Research Collaboration Effectiveness

For this chapter we note (and have experienced firsthand) that for most academics authorship, coauthorship, credit for contributions to scholarship, and other related criteria are important for career as well and personal reasons. Decision-making is therefore critical in research collaboration, and the norms for when to collaborate or not are equally as important. For example, who gets the authorship or coauthorship credit is important. Were the rules for contribution decided in advance or after the fact? Are my collaborators giving me proper credit for my contributions?

As decision-making analysts have known for many years, often the decision-making process is the primary determinant of outcomes (Brockner and Wiesenfeld 1996). While there is remarkably little evidence about research collaboration decision-making processes and norms, most agree (Katz and Martin 1997; Melin 2000) that these vital processes affect development of science in many ways. Included are researcher career trajectories and advancement, coauthorship decisions (Heffner 1981), choice of topics for study, and how and when to expend time and energy on a research agenda (Schut et al 2014). The primary issue seems to be the balance of benefits and costs of the collaboration and the factors that govern the quality of the collaboration experience and subsequent outcomes.

One relatively recent piece by Mayrose and Freilich (2015) indicates that shared research interests may factor heavily into decisions to collaborate and cooperate. There is an interesting dichotomy. Where there is significant overlap in research interests, fierce competition may exist, thus reducing the likelihood of collaboration. And where there is little overlap in research interests, the scientists may be challenged to communicate, also reducing the decision to collaborate. A more interesting finding perhaps is that despite the encouraging development of virtual collaborations using information technology resources, the authors found that social accessibility drives decisions to collaborate. Like the authors, we note that as systematic approaches improve for screening large- scale scientific repositories linked to data on affiliations of scientists, additional insights into collaboration decision-making will evolve.

Consistent with recent work in studies of interdisciplinary research centers (Bishop et al. 2014), we also note that case studies are important for understanding collaboration behaviors and effective decision-making in research collaborations. Specifically for this study, Bishop and colleagues found that decision-making in collaborative groups was improved by organized leadership, a positive research collaboration atmosphere, small group work, and the resources needed to collaborate with researchers outside of the current workplace or center.

This discussion is used to create a section, "Norms and Decision-Making Processes in Research Collaborations," in the propositional table at the end of the chapter.

How People Work in Research Groups

Collaboration entails the bringing together of knowledge, skills, abilities, and talents of researchers for the purpose of knowledge creation. However, that bringing together requires no direct or person-to-person interaction, meaning that all interactions could be remote or virtual by nature. Increasingly, very large teams of specialists produce research and publications, and, in some cases at least, some of the collaborators never meet in person or even interact with one another (Vanchieri et al. 2013). But we still accept that, virtual or not, research groups are the essence of collaborations, and research group work mostly involves people working with people. As such there are inherent interpersonal problems and issues with research collaborations.

Beaver (2001) has some of the best-known work in this area, offering, in his literature review article, a comprehensive examination of people in research collaborations. Among other aspects of collaboration, Beaver focuses on research collaboration processes, including feedback, dissemination, recognition and visibility, all of which he views as advantages of collaboration between collaborators. He argues for "synergy" among people in research teams such that multiple viewpoints of the collaborators enhance the project outcomes, including the less powerful, a prescription easier in small groups than in large hierarchical research teams (Considine et al 2014). That approach favored by Beaver is, of course, very much in line with the Consultative Collaboration Management approach we outlined in chapter 1 and that we examine in detail in the final chapter of this book.

Chompalov and colleagues (2002) provide some especially useful findings about the impact of human relations management styles in research collaboration. They identify four basic structures for collaboration, ranging from bureaucratic to participatory in nature. Bureaucratic collaborations are most successful when the project involves multiple organizations, and there must be a clear hierarchy to ensure that no one organization's interests are disproportionately served. Participatory collaborations are used when the outcomes of the research are the least formalized and have less differentiated structures.

Another interesting perspective on people in collaborations comes from Huang (2014), who finds that collaboration networks do not necessarily have a positive effect on research productivity outcomes. Homophily in research interests among researchers is, according to Huang, an especially popular approach but not necessarily the best approach; heterophilous communications and maintaining degrees of heterophily among people in a collaboration can be difficult and challenging but sometimes productive.

Much of the work in the science of team science genre focuses on human relations and human resources in research collaboration. The preponderance of articles in the team science literature focus on the adoption of small group findings from other contexts (e.g., management and sociology) and advocate application in scientific teams (e.g., Fiore 2008; Salazar et al. 2012) but do not provide empirical research findings about teams. Other studies (e.g., Pennington et al. 2013) provide or advocate frameworks for collaboration, especially multidisciplinary collaboration. Some studies provide analytical tools useful for collaborative teams. Thus, Mâsse and colleagues (2008) provide psychometric tools for assessing the integration of teams, focusing on, among other factors, trust and satisfaction.

In general, the team science literature at this point in time provides concepts, frameworks, and propositions and seeks to integrate literature, but it has not yet produced much empirical work on the operation and dynamics of scientific teams. To date the minority of team science studies that have presented empirical data more often provide qualitative data, including from interviews, focus groups, and ethnographic methods (e.g., Lotrecchiano 2013). The exception is work by Crowston and colleagues (2015), who examine data and interactions from the Data Observation Network for Earth (Data ONE) project. As the team science literature develops, it will be interesting to compare and contrast the empirical methods employed by scholars who tend to be more focused on approaches used in psychology and medicine with those used by the research collaboration scholars, who have historically employed methods more commonly found in management and sociology. For example, we might expect to see more work with team experiments and less with the surveys and bibliometrics tools that have dominated much of the research collaboration literature.

Gender Dynamics in Research Collaborations

Gender identity is obviously one of the most personal and salient issues in one's life, no less to academic researchers than to anyone else (Corley and Gaughan 2005; Gaughan and Corley 2010; Abramo et al. 2013; Johnson and Bozeman 2012; Pollak and Niemann 1998; Liao 2011). Understandably, given its importance in so many social and work-related domains, gender is sometimes an important factor affecting collaborative research teams. Etzkowitz and colleagues (2000) contend that two worlds of science still exist, one for male scientists and the other for female scientists.

A study by Long (2001) shows that from 1970 to 1995 there were significant advances in the entry of women into science and engineering. Yet a woman's ability to finance her education, as well as the salary differences between men and women academics, may inhibit academic careers of women and thus their ability to engage in research collaborations. This situation is not unique to science, technology, engineering, and mathematics, as women in the life sciences also are making strides in career advancement, but they are still lagging behind men in opportunities for academic contributions (Rotbart et. al. 2012).

Much of the work of Bozeman and colleagues (Bozeman and Corley 2004; Bozeman and Gaughan 2007, 2011; Gaughan and Bozeman 2016) focuses specifically on the effects of gender on research collaboration. Bozeman and Corley (2004) contend that research collaboration is a result of personal attributes of the collaborators themselves, gender not least. They constructed five regression models to examine collaboration patterns among academic scientists. One of these models analyzed the impact of tenure, grants, gender, and field on the percentage of female collaborators of an individual scientist, and according to the authors, findings were that "female researchers who hold the rank of non-tenure track faculty, research faculty, tenure track faculty, research group leader or tenured faculty collaborate with a higher percentage of other females than male researchers in the same ranks do" (607). Furthermore, Bozeman and Corley go on to say, "Especially noteworthy is the extent to which non-tenure track females collaborate with other females (83.33 percent)" (607). Evidence from these findings supports the idea that research collaboration patterns can vary greatly by gender.

In a more recent study, with its straightforward question, "How do men and women differ in research collaborations?" Bozeman and Gaughan (2011) examine gender as their primary focus in research collaboration, seeking explicitly to determine whether previously observed differences in men's and women's collaboration patterns are owing to actual differences in gender or to spurious associations related more to poorly specified statistical models than to actual differences (such as, for example, the fact that in most samples of academic researchers women tend to be younger than men, and models not allowing for this can distort results). Having developed a new survey questionnaire database under the US National Survey of Academic Scientists, data including more than 1,700 respondents weighted by field and by gender, the study focuses specifically on research collaborations with industry and on research collaboration strategies and motivations. The authors find that men and women differed considerably in their collaboration strategies, with men being more oriented to collaborations based on previous experiences. The Bozeman and Gaughan study was the first to give evidence of women as slightly more likely than men to engage in collaboration (at least if one controls for age and scientific field).

Other literature shows that traditional gender patterns in research collaboration seem to be changing. Van Rijnsoever and Hessels (2011), in their study comparing disciplinary and interdisciplinary collaboration patterns, find that women are more likely than men to engage in interdisciplinary collaborations. However, the findings must be treated with caution, inasmuch as they are based on survey data from a single university in the Netherlands with a low survey response rate.

Abramo and colleagues (2013) note that scientific advances require ever more collaborative efforts, and women seem to have a greater propensity for, and capacity to collaborate in, intramural and extramural research opportunities as compared to men. These findings must be treated with caution, inasmuch as only Italian women academics are studied using bibliometric or meta-analytic approaches.

If we use our broad-brush approach to describe research collaboration (meaning not just scholarly publications), then we find that women and men differ with respect to their patterns of collaboration with industry. Bozeman and Gaughan (2011) employ the "industrial involvement index" (Lin and Bozeman 2006; Bozeman and Gaughan 2007; Gaughan and Corley 2010; Ponomariov and Boardman 2008, 2010) to compare men's and women's collaboration with industry. The industrial involvement index is a weighted gradient (see Bozeman and Gaughan 2007 for detailed explanation) that aggregates a variety of types of interaction, ranging from modest and low effort (e.g., providing research papers upon request) to intensive (e.g., co-development of patents). Bozeman and Gaughan find that, even in a more fully specified model, men continue to be more involved with industry but that women's affiliation with multidisciplinary research centers tended to mitigate the effect.

These same authors (Gaughan and Bozeman 2016) examine gender dynamics in research collaboration based on interviews with 60 US academic scientists and 177 responses to a survey questionnaire. They find that gender structures important parts of collaboration but that status hierarchy and career stage, both of which interact with gender, tend to play a more important role in determining both activities and specialization in collaborative teams.

Institutional and Professional Contexts of Research Collaborations

What is considered effective research collaboration is clearly institution and context specific. We approach the idea of institutional and professional context of research collaborations from two perspectives. The first perspective is that collaboration is effective and productive due to institutional or professional context *processes and procedures* (Lee 2000). The second perspective is that the collaboration is effective because it is *ethical* per the norms of the institution or profession.

We start with Lee and Bozeman's study (2005), which uses a sample of 443 research scientists at university research centers. The authors find significant field effects for productive scientific collaboration. They control for field by identifying researchers as being in either "basic" or "applied" disciplines where the basic ones include physics, chemistry, and biology, whereas "applied" disciplines chiefly include all of engineering. The authors find a significant and positive relationship for applied scientists and research collaboration productivity but no comparable relationship for basic scientists. The study also points to one factor of considerable consequence to those wishing to evaluate productivity: while having more collaborators certainly increases citations during one's career, at least by a "normal count" (simply the number of times a person is cited), when using a "fractional count" (dividing citations by the number of coauthors), then collaboration has no positive effect on citations.

Audretsch and colleagues (2002) provide a discussion of academics at research universities as effective collaboration partners for private industry. They show that universities with network ties to firms tend to have greater research and development productivity, mostly due to firm access to the human capital from faculty and students at universities. The outcomes suggest that institutional and professional context does matter. Private industry network ties with universities result in research collaboration effectiveness and productivity.

Political and managerial priorities vary by discipline and profession (Hackett 2005), and this can affect research productivity. Changes in commercialization policies of universities affect productivity and collaboration patterns and seem to alter research agendas to some degree (Glenna et al. 2011). University and industry research structures and cultures are different, as are collaborations that involve only academic scientists and those that involve more complex university-industry partnerships (Lee and Bozeman 2005).

In addition to professional and disciplinary impacts, some literature is available that addresses organizational contexts and research effectiveness and productivity. Katz (2000) specifically addresses this concern. He argues that conventional measures used to evaluate research do not account for the nonlinear relationship between the size institutions associated with collaborations and the research performance. He goes further to argue that traditional measures of size and performance result in an exponential power-law relationship between size of the research group, institution or nation, and the perceived research performance. Katz makes his argument based on a variety of performance-related measures, including number of published papers, number of citations to papers, citations per paper, and number of coauthored papers (24).

Katz examines the effects of the size of institutions and the propensity to collaborate among university faculty and finds that "smaller educational institutions have a greater propensity than larger ones to collaborate domestically, particularly with industrial partners and other educational institutions" (29). He also reports that larger institutions are more likely to engender collaborations, both internal ones and international collaborations, and that collaborations tend to exhibit linear increases as the size of institutions increase. Katz's (2000) findings may be explained by the fact that researchers at large institutions generally have more capacity and human capital close at hand and need not go far to collaborate. By the same token, the propensity for researchers in smaller institutions to collaborate domestically with industry partners and other academic institutions more than likely has to do with the resource limitations of smaller institutions.

Turning to normative differences among institutions and disciplines, there are several useful studies of ethical issues in collaborations. During the past decade or so, researchers, especially those in the biomedical sciences (e.g., Rennie 1994, 2000, 2001; Wainwright et al. 2006; Cohen et al. 2004), have begun to focus on ethical issues and the "dark side" of collaboration. In various sections of this book we consider some of these issues related to honorary and ghost authors, among other problems, and in the ensuing section we examine the literature on "contributorship" policies and proposals that have been developed to address these problems.

Outside biomedical fields, research on the ethics and sociopolitical dynamics of scientific collaboration for specific institutions or professions remains scarce (Shrum et al. 2001, 2007). Perhaps this scarcity is owing to the view that such problems are neither as pervasive nor as troublesome in other science and technology fields as they are in biomedical research. To be sure, biomedical research has a different institutional and professional context than is found in other science disciplines. Medical research has special hazards resulting from unethical behavior, in part because of its massive operation of clinical trials (Devine et. al 2005; Klingensmith and Anderson 2006). Similarly, medical researchers can have strong ties to pharmaceutical industry, sometimes with troubling results (Diller 2005; Insel 2010; Elliott 2014).

Ethical issues in collaboration certainly are not restricted to biomedical fields, though there are important differences in problems and the intensity of these in various fields and disciplines (Chompalov and Shrum 1999). This is not an insignificant issue, as the challenges of researching complex topics are resulting in more transdisciplinary research collaboration, between research and industry and between different disciplinary fields and organizational contexts (Harris and Lyon 2013). What it means to be ethical or trustworthy comes from profession-specific norms, traditions, and history. The emphasis on commercial value as a byproduct of the collaboration is also related to norms and traditions of disciplinary fields (Ambos et al. 2008).

One important reason that fields and disciplines differ in their ethical problems of collaboration (or in the worst cases, collusion) is that the stakes differ. Collaborating social scientists do important work, but it is very different work, and the stakes are different than is the case for medical field trials or tests of the efficacy of pharmaceuticals. Similarly, research conducted by civil engineers, if incorrect or corrupt, can lead to flaws in public works, whereas any such problems in astronomy are much less likely to have immediate and dire consequences.

Contributorship Issues in Research Collaborations

Authorship on a collaborative research work is sometimes a power struggle, not surprisingly given the growing demands for academic achievement by researchers. It can be deceptively simple, indicating which contributors to include and in what order. For example, it is generally accepted that the scientists who make the greatest contribution should be included on the final knowledge product, but it is not always clear how this process of deciding contributorship affects other team members. We can expect from Merton's (1968, 1995) classic work on the "Matthew Effect" that credit will inevitably be disproportionate to more senior researchers, regardless of the particular nature or extent of their contribution compared to less well-known collaborators. One would expect that the collaborators and coauthors who receive less recognition from a given coauthorship would in some cases feel exploited, especially in those instances where they perceive their own contribution to be more significant than that of a more senior and well-known researcher.

Discussion of contributorship issues are often associated with health sciences researchers. For example, Levsky and colleagues (2007) describe a number of potentially troubling trends in authorship in medical journals between 1995 to 2005, including honorary authorship, ghost authorship, duplicate and redundant publications, and, most important, authors' refusal to accept responsibility for their articles despite their readiness to accept credit for professional purposes. They note that causes of the trends continue to be unknown, but that the relationship between authorship and career pressures is clear in the health sciences as well as in other science disciplines.

Researchers have considerable autonomy in their collaboration choices and collaboration strategies, and these are based in part on judgments about the conferring of coauthorship and status (Heffner 1981). The issue is who decides. Drummond Rennie, a deputy editor of the *Journal of the American Medical Association*, and a strong proponent of collaboration policies, acknowledged this deficit in a well-known editorial piece colorfully subtitled, "Guests, Ghosts, Grafters, and the Two-sided Coin" (Rennie and Flanagin 1994). Rennie is thought to have evolved the term *contributorship* to refer to the process in which authors declare in detail, usually at time of submission, their individual contributions to scholarly papers in the spirit of scientific transparency (Rennie 2001, 1274). Following a series of articles that describe a growing problem with irresponsible authorship of medical research articles, Rennie proposed a major change in instructions to authors contributing to the *Journal of the American Medical Association* (2000, 89). These changes in contributorship requirements have provided clear signals where none were given before and, presumably, have enhanced collaborators' ability to communicate effectively with one another about contribution and credit. However, even in journals adopting contributorship policies, we still know little about the *validity* of contributorship statements or the social and potential power dynamics entailed in developing them. To date, too little research systematically assesses the effects of coauthorship in general and contributorship norms in particular, despite the fact that such standards have been widely adopted in medical and health sciences fields. Indeed most studies of practices of authorship come from the biomedical and health sciences fields (Marušić et al. 2011), in part due to the nature of multidisciplinary clinical research.

Beyond the biomedical disciplines, there do not seem to be many empirical studies on contributorship, though some systematic approaches to studying the contributorship are evolving. Youtie and Bozeman (2014), for example, find that "bad collaborations" can range from miscommunication to clear exploitation and legal disputes about contributorship. Having multiple institutions involved in a collaboration can improve the odds of a productive and good relationship between contributors to a knowledge product, but, again, the varying norms and standards for decision-making take time to sort out, and this conversation is often best had early in the collaborative process. For example, honorary authorship is fairly widespread in the biomedical sciences but can range from being discouraged in other disciplines to being tagged as an unethical authorship practice (Moffatt 2011).

As mentioned above, some "dark side" aspects of collaboration have received very little attention. While there is widespread concern about the possibilities for student exploitations in collaborations where students do not receive credit for their contribution to knowledge products (e.g., Slaughter et al. 2002), most of the evidence thus far is anecdotal or unsystematic. The few systematic case studies (e.g., Baldini 2008) available suggest problems but give no clues about the extensiveness of student exploitation in collaborations. Moreover, there is some evidence that collaborations rooted in industry-university partnerships often have beneficial effects for students, including early publication, job offers, and mentoring (Welsh et al. 2008; Bozeman and Boardman 2013). Dooley and Kenny (2015) find that students are quite receptive to working with industry and are interested in commercial aspects of work, but at the same time they strongly prefer a balance between traditional disciplinary work and commercially driven work.

More empirical studies are needed to sort out the various contributorship dynamics and the possible negative consequences of contributorship disputes on knowledge production. This discussion is used to create a "contributorship" section in the propositional literature table at the end of the chapter.

Promoting Better Research Collaborations

Our book is about making research collaboration and collaborative teams more effective, and much of the existing research literature is quite relevant to this objective. First, we note that the number of studies employing different data and methods have provided evidence that collaboration tends to enhance productivity of scientific knowledge (Pravdić and Oluić-Vuković 1986; Lee and Bozeman 2005; Wuchty et al. 2007; Huang 2014). In the case of collaboration's effects on profits, wealth, and economic development, the models tend to be more complex, but here too the preponderance of evidence is that research collaboration has beneficial effects with respect to scientific contributions (Franklin et al. 2001; Shane 2004; Dietz and Bozeman 2005; Link and Siegel 2005; Perkmann and Walsh 2009).

So if research collaboration enhances productivity, or, to put it another way, if researchers when working together give rise to greater contributions to science knowledge, then there are good and practical reasons for describing, organizing, and designing the state-of-the-art collaboration network. Countless resources and human energies are invested in facilitating, inducing, and managing collaborations (Allen 1977; Hagedoorn et al. 2000; Sonnewald 2007), and thus the question is not whether research collaboration provides benefits but whether those benefits are sufficient to warrant the prodigious investment of resources. In addition to the "is it worth it?" question, it is certainly the case that some collaborations are highly productive and others less so, ergo, the "what works best?" question.

Many studies focus on industry-university relations and find consistently that research collaboration with private industry is driven to a large degree not by university requirements or pressures but because researchers work in these firms. We know that the objectives, composition, and content of research in industry tend to be quite different from those found in universities, government, or nongovernmental organizations (Crow and Bozeman 1998; Cohen et al. 2002; Guellec and Van Pottelsberghe de la Potterie 2004; Mitchell et al. 2015). Thus there is much to be learned from literature about research in private industry, though we know it mostly draws from research traditionally produced in universities (Mansfield 1995).

One recent article by Pohl and colleagues (2015) describes the factors that promote interdisciplinary research collaboration productivity linking academic to private industry experiences. With a focus on publications, and using qualitative research methods, the authors identified groups of resources that determine research collaboration effectiveness. The categories include (1) the focus on scientific research collaborations; (2) the coordination of research teams; (3) availability of feedback for the research products; and (4) a chronology of research collaboration success. Further, and according to this analysis, effective research collaborations will include both shared and open-minded vision by collaborators, as well as well-defined collaboration management structures and principles.

We also see another area of instructive literature, drawn from business literature, concerning academic entrepreneurship and the idea that entrepreneurship focuses on measurably effective outcomes. Though not without its ongoing critics (Slaughter and Leslie 1997; Rhoades and Slaughter 1997), we see growing interest in property-focused academic entrepreneurship and associated research collaborations. A great deal of the literature on academic entrepreneurship considers the impacts and practical uses of knowledge-focused research, particularly entrepreneurship resulting from research collaborations. The chief arguments against academic capitalism are that (1) industrial involvement has unduly affected university researchers' choice of research topics and, perhaps more important (Mendoza 2007; Cooper 2009), that (2) this has led to an exploitation of graduate students, who have become "tokens of exchange between academe and industry" (Slaughter et al. 2002, 282). Still, it is a concept worthy of continued discussion and exploration when considering strategies for research collaboration effectiveness.

The academic entrepreneurship and capitalism literature is actually much larger than its research collaboration component, and we address here very little of the academic entrepreneurship literature, only those aspects that related directly to collaboration or collaborative teams. But at least some of the technology transfer component of the entrepreneurship literature has implications for collaboration (e.g., McKelvey et al. 2015; Cassi and Plunket 2015).

In a series of related studies, Siegel et al. (2003, 2004) and Siegel, Waldman, and Link (2003) provide broad-based empirical analyses of university technology transfer offices that indicate a number of ways in which these offices can be either a help or hindrance in collaboration. Siegel et al. (2003) report results from interviews with university administrators, scientists, and business professionals. Their recommendations are specifically targeting the technology transfer offices at institutions, often the source of productive research collaborations, and they include basic business management principles such as (1) understanding the collaboration partner needs; (2) having flexible agreements; (3) considering rewards and incentives for partners; (4) having adequate resources to bring to the collaboration; (4) and valuing the personal relationships and social networks between collaboration partners.

These findings are consistent with other studies (Renault 2006) that find that institutional policies can negatively influence collaborations between university and industry researchers. Even in instances where higher-level university administration is committed to industrial partnership, middle and lower levels of bureaucracy sometimes sabotage these goals (Audretsch 2002). Nevertheless, industry outcomes often are net positive for university interactions, and that explains the continued focus on varied forms of industry-academia research collaborations (Behrens and Gray 2001).

Conclusion

This book uses many sources and approaches to explore research collaboration effectiveness, and the existing literature, developed for many different purposes by people in a multiplicity of disciplines and fields, is often useful for those wishing to find lessons to apply and improve collaboration. This review and the related propositional inventory in the appendix are extensive but certainly not comprehensive. For those interested in more complete reviews of the relevant literature, there are many sources available (Melin and Persson 1996; Katz and Martin 1997; Melin 2000; Bozeman and Boardman 2014; Leahey 2016), all with useful lessons about collaboration effectiveness and "what works best." Moreover, throughout this book we try to employ the literature as we discuss particular problems of research collaboration and ways to improve collaboration.

4

Thinking Systematically about Research Effectiveness

Introduction

Since our book is not an abstruse rumination about research collaboration but rather an instrumentally focused book aimed at improving collaboration, simple concepts that give some purchase on effectiveness seem to us well justified. Thus, we speak of such as Nightmare or Dream Collaborations, the Tolstoy Principle, and so forth. Still, the book embraces the simple but not the oversimplified. We offer no panaceas, bromides, or Top Ten lists, nor do we wish to ignore or gloss over the contingencies, complexities, and overall messiness inherent in research collaboration. Simplicity is a good starting point, but improving the effectiveness of collaborative teams ultimately requires attention to their complexity.

With this chapter, complexity's time has come. We enter in with full understanding of the need to avoid depicting research collaboration as a chaotic mix of unpredictable human experience. Indeed, we shall see that much that occurs in research collaborations is predictable, at least in broad outline. One hoping to improve research collaboration effectiveness must keep an eye out for patterns and must understand their causes. It is not sufficient to say that "physics is different," or "hyper collaboration relates not at all to small number collaboration," or "industry collaborators change everything." We have to look for patterns among the particularism.

Our complexity management strategy is to present a model we refer to as the Aggregate Model of Research Collaboration Effectiveness and to then apply the model in reference to our qualitative data. Why an "aggregate" model? To this point we have found it convenient to ignore or de-emphasize two issues important in understanding research collaboration effectiveness: (1) the fact that there are usually multiple criteria for effectiveness; and (2) the fact that what is effective for one collaborator may be highly ineffective for another. In this chapter, we give considerable attention to these issues, in fact in the section immediately below. However, it is also worth noting, as we do in the Aggregate Model below, that the need for nuance does not undermine the need for an aggregate conceptualization. To put it another way, we want to answer this question: "Typically, how does factor X affect research collaboration outcomes?" not the question, "How does factor X affect physicist Sheila Smith's perception of effectiveness according the criterion Y?" Thus, what we are aggregating here is the multiple perspectives and multiple criteria. It is important to understand that these particularities exist and, at the same time, to avoid allowing them to paralyze our thinking about research collaboration effectiveness. Researchers have no problem thinking broadly about collaboration effectiveness. When we asked them to tell us about their effective or ineffective research collaboration experiences, not one of them paused to ask, "Which dimension of effectiveness?" or, "By what criteria?" However, their responses showed deep and particularistic thinking about multiple dimensions of research collaboration and its effectiveness, thinking consistent with the discussion below.

Objective and Subjective Conceptualizations of Research Effectiveness

In examining research collaboration effectiveness, a first question to consider is this: "Why not just focus on the tangible output, the articles produced, patents, and the citations?" There are many useful quantitative studies examining the effect of research collaboration on publications and other more or less objectively measured knowledge products. We like these studies, and not only because we have produced many of them. Quantitative studies of research collaboration (see Bozeman and Boardman [2014] for an overview) have the great advantage of precision. There is not much room for opinion, and the issues are not so squishy. People collaborate and produce measurable goods that have measurable outputs and impacts (albeit some more measureable than others). Still, the more precise quantitative studies do not tell the whole story in terms of the impact of experience on subsequent choices, including choices about whether to collaborate, with whom to collaborate, and how (processes, structure) to collaborate. We know from previous research, both quantitative and qualitative (Melin 2000; Shrum et al. 2001, 2007; Youtie and Bozeman 2014), that research collaboration decision-making is sometimes quite complex and multifaceted, and it includes many and diverse considerations. Factors and motivations that may be extremely important to one partner in a collaboration may be much less important to another.

In this study we spend little time on directly measured output. Much of the data we present, though certainly not all, is subjective, and we feel subjective data have great importance, not least because the experiences of collaboration participants are experienced subjectively. Our concepts and indicators are not invariant, inasmuch as the interviewees provide form, shape and depth to notions about research collaboration effectiveness. We feel this approach, while certainly not the only possible useful approach, has the advantage of recognizing that researchers look for different things in collaborations and collaborators (Bozeman and Corley 2004). What is a good collaboration to one researcher may be suboptimal to another (Katz and Martin 1997; Melin 2000).

In our view, knowledge of the relationship of collaboration to tangible outputs such as publications and citations is very useful but is not enough. Here are some of the reasons why. First, when collaborators make contemporaneous judgments about collaboration effectiveness, publications outcomes are almost never among the "earliest returns." We know from evidence we present here that collaborators do not suspend judgment about the effectiveness of collaboration or the value of collaborators while they await acceptance of research for publication or the accumulation of citations. Since early judgments have strong psychological framing effects, these early judgments necessarily affect later decisions, even in the face of new or more telling evidence (Mandler 1980; Herr 1986; Muchnik et al. 2013). Second, even after a publication is completed, it may take a good deal of time to determine the impact of the publication. Third, and relatedly, the fact of publication does not necessarily give insight into the collaborators' assessment about whether the work achieved its potential (e.g., "Was the publication in the best possible journal?" "Was the impact as great as expected?"). Fourth, we know from previous studies that while publication (or patenting) is important for almost all collaborators, it is typically only one among several motives for collaboration. Other motives include mentoring and developing scientific and technical human capital (Bozeman and Corley 2004); obtaining a stream of research support (Bozeman and Gaughan 2011); or contributing to institution building (Ponomariov and Boardman 2010).

In short, regarding the discrete and observable research outputs such as publication or patenting or citations, all of these are important, but so are the subjective judgments of the collaborators. Moreover, our approach seems to have some validity, as evidenced by the fact that our respondents did not have to grasp for straws while telling us about their good and bad collaborations; it was therefore not difficult to identify good and bad collaborations from the evidence provided. Respondents described each with no difficulty whatsoever and, to some extent, with common themes and content.

An Aggregate Model of Research Collaboration Effectiveness

As is the case for most research areas, research on collaboration and team science focus unevenly on the phenomena, with some presumed causes and effects receiving great attention from researchers, others less. Thus, previous research has focused extensively on such topics as strategies and motives for collaboration (Bozeman and Corley 2004; van Rijnsoever and Hessels 2011; Tartari and Breschi 2012), effects of collaboration on research productivity (Gaughan and Ponomariov 2008; Lee and Bozeman 2005; Cummings et al. 2013), effects of geographic proximity on collaboration (Katz 1994; Olson and Olson 2000), effects of new media technology on collaboration (Finholt 2002; Powell et al. 2012); and developing professional guidelines for documenting collaboration contributorship (Madden 1998; Frankel and Bird 2003; Jones 2003; Austin et al. 2012). Almost all these issues pertain to the quality and effectiveness of research collaboration outcome, but so do others that have received less attention until quite recently (Leahey 2016).

Figure 4.1 depicts the Aggregate Model of Research Collaboration Effectiveness. As suggested above, many empirical studies of research collaboration have been conducted, and from these we can identify factors related to research collaboration effectiveness. Some research focuses on issues outside the collaboration but significantly affecting the collaboration. Thus, the model depicts the major External Factors related to research collaboration effectiveness, including issues related to disciplines and field specializations, factors related to interdisciplinary research and interdisciplinary teams and the interaction of institutions, including, but not limited to, the interaction of universities and industry in pursuit of commercial goals.

The model also includes Internal Factors, issues related to the specific collaboration and characteristics of collaborations. Some of these are "team science" factors, but others relate more to the characteristics of individual collaborators. Thus, the Internal Factors include issues related to the fit of work style, the collaborators' previous experiences with one another, and, relatedly, the amount of trust among collaborators, the career stages of the collaborators, and the gender of the collaborators. Each of these Internal Factors is examined extensively in the research literature.

The model also shows the category "Collaboration Management and Processes." The research literature on research collaboration has paid remarkably little attention to issues related to the management of collaborations, despite its apparent importance. The science of team science approach often discusses the importance of processes and management (e.g., Cooke and Hilton 2015) but has yet to produce much evidence on the topic. It is surprising that both the general research collaboration literature and the science of team science literature include so few studies of collaborative team management. One might anticipate not only that management is important to any group enterprise, even one that is often so fluid as research collaboration. This may especially be the case when we consider that persons who have no management training of any sort are in charge of most research collaborations and that they tend to, therefore, embrace a top-down

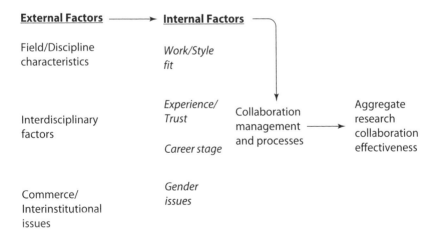

FIGURE 4.1. Model of *aggregate* research collaboration effectiveness.

management approach that comports with their experience as principal investigators (Bozeman and Rogers 2001). We return to this point in the final chapter's discussion of Consultative Research Collaboration.

DISCIPLINARY AND INTERDISCIPLINARY DYNAMICS

Many aspects of collaboration are sensitive to differences in scientific fields (Ennis 1992, 2000; Chompalov, Genuth, and Shrum 2002; Bozeman and Boardman 2013a. This is not only due to the nature of the technical work itself but also to such factors as differing work norms, differing styles of graduate education, varied types and degrees of specialization, and whether it is even possible to realistically conduct research as individuals as opposed to teams. Some fields that virtually *require* collaboration, due to the sociotechnical organization of work, are somewhat more likely to show collaboration effectiveness simply because individuals in such fields will have greater experience with collaboration and because their careers depend fundamentally on their ability to work effectively in teams. Relatedly, fields such as mathematics or most fields of social science often have worse outcomes due to lesser degrees of experience and lesser degrees of institutionalization of practice.

Any approach to research collaboration or team science that fails to grapple with the differences among fields and disciplines misses much. In extreme cases, behaviors that are considered unethical in some fields are considered normal practice in others. Here is an example. One of the economists in our study related a case in which a visiting international doctoral student from a different discipline said to him, in a passing comment, "I hear from one of your doctoral students that you do your own research work." The economist, startled, said, "Of course I do my own research, if not, who else would do it?" The student noted that in his lab group the professors rarely collected data or even wrote up results. Rather, they worked with students to develop very rough ideas, the students went away and did the lab work, gathered the data, wrote up the paper and put the adviser's name on it, sometimes as first author. The economist was horrified at an "unethical" practice considered routine in some lab-based, PI-dominated fields.

COMMERCE AND INTERINSTITUTIONAL FACTORS

We expect that collaborations among people in the same university in the same department in the same laboratory are, all else equal, less likely to be fraught with bad outcomes. When collaborations involve multiple departments, or multiple universities, or personnel within universities, or even the very different work settings of industry and government laboratories, matters get complicated. Boardman and Bozeman (2007), in their study of university research center participants, find that likelihood of bad collaborations increases almost in a linear function with the degree of institutional heterogeneity (Boardman and Bozeman 2007). However, Youtie and Bozeman (2014) found an inverse relationship between the share of authors at different universities and the likelihood of problems, which suggests that sometimes being at the same university offers greater convenience for having disagreements.

A particular variety of interinstitutional collaboration that seems to have potential for peril to both parties is university research collaborations with industry. When commerce enters, conflict often (though not inevitably) ensues. The research literature on industry-university relations warns us of the hazards of such collaborations (though it also documents the many benefits). At risk of anticlimax, we note here that our own findings, some of them reported in the next chapter, show only limited evidence of commercial transactions being a major impediment to research collaboration effectiveness. The sky-is-falling tone of some of the literature on university-industry research collaboration is reflected in none of our data. To be sure, problems were reported but were moderate in tone, focusing on frustrations in publishing delays or in dealing with collaborators whose goals were sometimes not in alignment. In most cases, the frustrations were not sufficient to cause researchers to drop out of collaborations or even to avoid collaborating in the future with the same parties.

One respondent acknowledged that standard commercial management decisions concerning market size and mergers and acquisitions limited the ability of an academic scholar to collaborate with the private sector on medium- and long-term research activities:

> I got approached by [a company] that builds equipment for [a subfield]. They were thinking about building a next generation amplifier that could do the [specific technique]. They wanted me to be a consultant and they started sending me the first drafts as a user to give feedback. And this went on for half a year; then they got bought by [another company] which decided that the market wasn't big enough and it was scrapped.

Perhaps one reason that we have few reports of major problems with university-industry commerce-based clashes is that many of the more serious problems are not at the level of the individual scientists (our focus) but rather at the level of the research vice presidents and others grappling with more general intellectual property issues.

WORK STYLE FIT

In research collaboration, as in nearly any work enterprise, it is sometimes the case that colleagues lack "chemistry," that even if they are each a competent and effective worker, their styles, personalities, and work habits are just not compatible. Where there is a poor fit of work styles, research collaboration effectiveness suffers.

Since work style is a broad concept, it is useful to give a few examples. Often people who work at a steady pace find it difficult to work harmoniously with people who work at a boom or bust pace. People who like to "talk out" research problems sometimes do not mix well with people who like to come to initial agreements and then work independently on jointly identified issues.

Work style problems can come in many forms, but here is a fairly typical instance of a work-fit collaboration problem, not a Nightmare Collaboration but a Routinely Bad research collaboration owing to poor work-style fit:

> I had a problem working with someone whose timing was way off compared to mine. We were doing the experiments and someone at University X was doing a small part of the experiments. They were so incredibly slow with doing the experiments and editing the paper. It was like pulling teeth getting anything out of them. I was hammering on the lead PI to get it done, get it done. She became very frustrated with me, just as I was very frustrated with her. We had very different ideas about work pace. She liked to do many things at once and I liked to do one thing quickly and then go to the next.

It is important to note that work style is not often or even usually a problem. Indeed, when asked about their best collaborations, one factor most frequently cited is work style. This seems an issue that almost all collaborators consider in constructing research collaborations, and thus problems are avoided.

CAREER STAGE

In many cases, colleagues recognize that individuals at different career stages have different objectives for the collaboration (Zucker 2012). In most cases it is easy enough to accommodate those differences in objectives, and there is evidence that senior colleagues often take much care in accommodating the interests of more vulnerable junior colleagues, postdocs and graduate students (e.g., Bozeman and Corley 2004). But in any collaboration where the career motives and the team members' perceptions of the career-related objectives for the collaboration begin to clash, the potential for bad collaborations is greatly exacerbated. Sometimes the career stage issues can be at the root of very bad behavior.

In general, the data we present in later chapters show that collaboration dynamics that are related to rank and status differences are sorted out by the collaborators themselves, in many cases with the junior collaborators expected to be accommodating. These experiences do lead to the development of career-stage heuristics and even abandoned careers, but it does beg the question of whether people leave the field as junior researchers because of power dynamics related to career stage or because of other disenchantments or opportunities.

TRUST AND EXPERIENCE

Some of the issues in research collaboration effectiveness seem very much related to the particular context of research and its social dynamics. For example, there are few counterparts in any other type of workplace collaboration to the intricate processes and career outcomes related to tenure. But other issues in research collaboration are not so different than one might expect from other forms of work collaboration, ones that have no relation at all to research. Trust is one of the most often cited factors in team building and collaboration effectiveness generally (Vangen and Huxham 2003; Inkpen and Currall 2004) and, thus, it should surprise no one that it is so often mentioned in connection with research collaboration effectiveness (Dodgson 1993; Baker 2015; Fior et al. 2015).

Consider the following typical account:

> My best collaboration is the one that has been going on for 12 years, or most of my postgraduate experience. It was a chance meeting and our skills complement each other. We are very comfortable and communication is very easy.

For the most part, trust and experience working with one another are factors usually mentioned together, and in positive terms. As we will see in a later chapter, a lack of trust or violations of trust are sometimes cited as major factors associated with bad collaborations.

GENDER DYNAMICS

Nearly all of the women we interviewed, and at least a few of the men, reported that gender issues had at some point negatively affected their research collaborations. In some cases these issues played out exactly as reported in the feminist literature (Hanson and Pratt 1995; Banks and Milestone 2011) on occupations and workplace dynamics, with nothing specific to STEM careers; in other cases the issues were more distinctive to academic science. In some cases, problems go directly to biology, especially childbearing. And sometimes the issues begin very early, especially when children arrive during graduate school:

> I was the first graduate student to have a baby and get my PhD—the first one. There weren't many women to begin with, and all of the other women, if they had a baby, they wrote a master's and left. I was really surprised because I had gotten through classes, I was a top student in terms of grades, had the top score for my public seminar, had done great on my advancement exam. After all that, I had two papers in press, and then I announced my pregnancy . . . then—not my advisor, but others—faculty said, oh, when are you writing your master's, are you going to leave? They just all assumed I was done.

While many of the issues related to gender are very directly related to gender—for example, issues related to child bearing—others are much more subtle and have to do with culture and social construction of gender. In academia, as in most of society, people bring with them to work their sometimes very different assumptions about the meaning and implications of gender. Often these get in the way of effective research collaboration. Complicating factors further, sometimes it is difficult to know whether gender issues are at the root of a problem or whether gender differences are incidental to a clash of personalities or goals.

COLLABORATION MANAGEMENT AND PROCESSES

Most collaborations have major decisions to be made, ones such as determining publication objectives, choosing presentation and publication outlets, and, especially, settling on approaches to assigning credit. For such decisions, good management is paramount. However, studies of the management of academic collaborations are almost nonexistent. There are likely several reasons for this. First, some researchers clearly do not think of collaborations as even being an object for management. They assume that collaborations are loosely coupled processes where all participants are volunteers and all are benefiting, and, interestingly (and incorrectly), they sometimes assume that there is an unspoken but unanimous agreement about inclusion, credit, and dissemination of findings. Second, when researchers hear "management" they often equate the term with "hierarchical management" and, thus, they feel that if they are being participative then no management is occurring. Third, most researchers simply do not think systematically about management and likely prefer not to do so. But as we shall see later in this book, having good management and effective group decision-making in collaborations is one of the best ways to avoid bad collaboration outcomes.

Collaboration management issues can create significant tensions, and, importantly, they are often amenable to simple resolution. Thus, teams that have early, frequent, and participative discussions of credit sharing and other strategic choices less often have bad collaboration outcomes (Youtie and Bozeman 2014). We discuss these issues in our concluding chapter, much of which is devoted to approaches to managing research collaboration.

Multiple Dimensions of Effectiveness

While the aggregate effectiveness conception has served us well—it is helpful to think of general issues of effectiveness—it is now time to consider more specific, more complex notions of research collaboration effectiveness. Even while making generalizations about research collaboration effectiveness, it is good to bear in mind that effectiveness is actually multidimensional. This is true not only of collaborative teams but also of all sorts of organizations.[1]

It is easy to understand that research collaborations have multiple dimensions of effectiveness. Let us consider an example. Let us say a particular collaboration turns out to have had highly positive scientific and technical results, perhaps enhancing the careers and reputations of coauthors, but results in collaborators disliking one another so intensely that they are no longer on speaking terms. Similarly, collaborations may have excellent commercial outcomes but at the same time frustrate parties to the collaboration who are aiming to have publications in referred journals (or vice versa).

In addition to being multidimensional, the effectiveness of research collaboration is in many respects perceptual, and perceptions sometimes differ radically. The first interview data presented in this book was the one we referred to in chapter 1 as the "Rock Star" collaboration, in which the respondent thought that this was an all-time best collaboration experience despite the fact that the Rock Star collaborator did very little work other than lending his name and taking credit as first author. It was a Dream Collaboration, from the standpoint of the respondent, because it resulted in a string of excellent, influential publications. As the respondent noted: "Even if Adam [aka Rock Star] doesn't do anything, a paper he is on will get published just because of his name. A proposal will get awarded just because of his name."

What this respondent views as his "best collaboration" involves behavior that some would view as unethical or at least unprofessional: claiming primary authorship credit when one has done little work. We have already discussed briefly the issue of honorary authors and the problems posed by the fact that researchers receive credit in cases where they have little or no involvement in the research. However, in the Rock Star case there is no implication that Adam failed to contribute to the paper; Adam is after all described as an "idea guy." Quite possibly one of his ideas was as important to the paper as any other contribution. In short, making judgments about collaboration practices can be complicated, and different people can come to different conclusions about what is an effective, acceptable, or ethical practice.

The line between collaboration categories is not always clear cut, and one reason is that collaboration effectiveness is multidimensional. Another reason is that effectiveness is situational, and the above case illustrates both the multidimensionality and the contingencies of effectiveness. Regarding multidimensionality, Adam the Rock Star did no direct work on the research but provided benefits from his accumulated social capital and, most important, was to some degree (we do not know to what degree) instrumental in the idea for the work. Thus, from the standpoint of affecting the ultimate outcome and impact of the research, it is possible that Adam proved to be an exceptionally effective collaborator, even if he proved unusually ineffective if one's criterion is actually helping to produce the research in its operational aspects. In different cases and from different collaborators' perspectives, this same experience could register as either a Dream or Nightmare Research Collaboration. Honorary authorship could qualify as a Nightmare Collaboration if such an outcome is one of a series, a pathological problem; if the intellectual property involved is especially valuable, career making for the person who actually did the work, then the result could be a Nightmare Collaboration; if a beginning researcher is exploited and shunted to the background in favor of the ghost author, the results could be long lasting and thus a Nightmare Collaboration. However, if the case is just an instance of a busy and harried person letting others do most of the work on a barely formed idea and then letting all exploit his hard-earned reputation, then not only is a Nightmare Collaboration avoided, but some might (as this respondent did) judge the collaboration as a highly effective one.

A final point on the Rock Star case: in interviewing several researchers who clearly qualify as star researchers, we discovered that it is not at all uncommon for such individuals to have requests to collaborate in which the person wishing to collaborate will say, essentially, "Don't worry about doing any work on the paper, you can just read and comment," proposing to trade their labor in order to benefit from the research star's name and social capital. Indeed, this could be a rationally self-interested strategy for the lesser-status person and an act of generosity on the part of the honorary author. That, incidentally, is one of the reasons why we suggest, in later chapters, that research effectiveness is not best viewed in terms of personal gain. Trading on status may be good for individual researchers but not so beneficial for the research enterprise.

Multiple Perspectives: Subject Dependence and Research Collaboration Effectiveness

In addition to there being multiple *dimensions* of collaboration effectiveness, often collaborations include multiple *perspectives* among the collaborators. This is well illustrated by the following comments, a response provided by a researcher asked about "especially bad collaborations":

> I was at [identifies home institution] and we had a collaboration in which people from UC-Davis and Berkeley were involved. At the end of the study I chose one of the collaborators to write a proposal with. George, the overall PI of the study[,] had involved Susan, who got pregnant and did nothing. Sam did all the proposal writing and there was a huge amount of follow up. Both Susan and Sam were told by George that they would be the PI on next project and eventually they were made co-leads. So I worked with Sam, but thought we should talk to Susan because it is ethically right. Susan said she also wanted to be involved and that all the work was her idea. We decided that it wasn't going to work out to involve Susan. Then she said "I will sue if you submit this." I was living on soft money at the time and needed the grant. So that was hard as I had to pull the grant from the Fedex mail drop. The dean got involved with another dean and it went all the way up to the Vice Chancellor of Research. The finding was that person Susan didn't have an argument.

We never did submit the grant (we submitted one related to it later) and that was painful. We did submit a later [grant] (without Susan) but it didn't get great reviews.

This is a complicated case. One could infer that there is a bad collaborator here, "Susan," who did little work and wanted to hang on, despite doing little or no work, delaying the project and leading to disputes that had to be resolved, to the extent they could be, by higher-level university administrators. That is one interpretation, quite possibly a valid one. But we have no interview data from Susan. Another possibility is that Susan developed an idea for a project (the interviewee suggests as much), became pregnant, and then like many academic women having babies either went on pregnancy leave or slowed down her work schedule. Doing so, presumably, would not mean passing off her idea for others to develop while not including her. Yes, this is perceived as a bad collaboration, but for whom?

Research Collaboration Effectiveness: A Multidimensional/Multiperspectival Concept

As we see, if we consider research collaboration effectiveness as something more multifaceted, the complications quickly arise. Thus, we present the figure below, which provides a multidimensional/multiperspectival conceptualization of research collaboration effectiveness.

One way of sorting out the complexities of collaboration effectiveness is to be explicit about component dimensions. Any research collaboration model is a bit closer to reality if it explicitly recognizes that outcomes may have different meanings and impacts for different parties to the collaboration (and for parties external to the collaboration). Taking these factors together, we have what we refer to as a multidimensional/multiperspectival model of research collaboration effectiveness. This is presented in figure 4.2 below.

Based on the types of factors our respondents cited when discussing the quality of their research collaborations, we offer the model depicted in figure 4.2. The focus of the model is on impacts for the individual participants and the corporate enterprise of the collaboration. However, there may be important external stakeholders. For example, the idea of organization support impacts is a very broad one. We can consider impacts of the collaboration on, for example, the laboratory or center with which the collaborators are associated, the academic department, or the university as a whole. Likewise, students who have not actually participated directly in the collaboration and who are not coauthors or stakeholders in any other sense may nonetheless

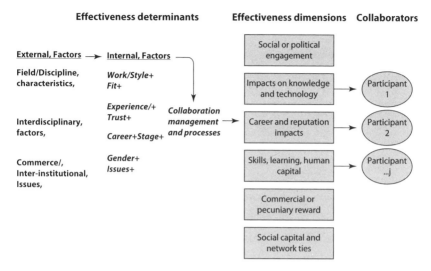

FIGURE 4.2. Multidimensional/multiperspectival model of collaboration effectiveness.

benefit if important aspects of the collaboration are included in classroom instruction or mentoring. Similarly, if students receive research funding as a result of the collaboration, that is an important impact but one not necessarily internalized by the collaboration.

There are ways of thinking about aggregate effectiveness while at the same time accommodating multiple perspectives. In the first place, one can maintain that aggregate means additive, that it is simply the total of all outcomes for all parties to the collaboration. In the second place, we can think of effectiveness as what the respondent means when they refer to "an especially good" (or "an especially bad" collaboration). Respondents doubtless differ to some degree in what they view as an effective collaboration, but most of them mention some combination of the factors provided in the model above with the most common ones being direct knowledge and technology impacts and career or reputational impacts. But each of the effectiveness components were mentioned by several of the respondents, and, indeed, more than 95 percent of all effectiveness factors identified by respondents would fit easily in one or more of these categories.[2]

EXTERNAL STAKEHOLDERS

One limitation of our study is that we do not pay much attention to the broadest implications of research collaboration on external stakeholders or the large-scale social impacts of research. As science and technology ultimately produce knowledge products and tangible products that are put into use, particular collaborations may affect entire fields of inquiry, organizations and institutions, and the individuals who may be users of products. These products may be "end use" products, such as a discrete technology developed from a patent based on a research collaboration. More often, research and technical work provides incremental impacts that contribute to a stream of work that ultimately has social and economic impact. In such cases, instances where there are a great many research products contributing to technical, economic, and social change, ferreting out and evaluating the specific contribution of any single research collaboration may be literally impossible.

The basic point: research collaborations *may* have impacts well beyond the persons who are collaborating. How frequently does that happen? We do not know. But most studies of "normal science" would seem to indicate that the typical knowledge product has no long-lasting effect for persons not a party to its production. However, even if this is the case, then it is difficult to trace butterfly effects. Perhaps an incremental study provides the increases in learning and human capital that permit one or more collaborators to produce a later work having much greater impact. Or it may be the case that a student learns valuable information from the knowledge product produced from the collaboration. Whether the student actually works on the research team or sits in another country far away and reads about the work, the study still has the potential to affect human capital, the store of knowledge, and, thus, to have impacts in ways difficult to trace.

Research Effectiveness and Research Ethics

A final important topic needs at least some consideration before leaving this chapter and proceeding to other chapters focused on empirical findings about research effectiveness. Our book is not particularly in the research ethics genre, but to some extent ethics impinges on effectiveness. Clearly ethical problems are not only matters of individual conscience, they can undermine collaborations, careers, and even the whole of research. .

During the past decade or so, researchers, especially those in the biomedical sciences, have begun to focus on ethical issues in collaboration (Rennie et al. 2000; Williams et al. 1998; Cohen 2004). Lagnado (2003a) argues that trust in the meaning of coauthorship has eroded. The relationship between authorship and career pressures on academic medical researchers is particularly acute, as manifested by new authorship criteria having been developed by the International Committee of Medical Journal Editors (ICMJE 1997; Jones 2003) in a policy considered a model for some professional groups and other science and engineering societies

Outside of biomedical fields, research on the ethics and sociopolitical dynamics of scientific collaboration remains scarce (Shrum et al. 2001, 2007). Perhaps this scarcity is owing to the view (we think mistaken) that such problems are neither as pervasive nor as troublesome in other STEM fields. To be sure, biomedical research is different. In most STEM fields there is little potential for unethical behavior to affect clinical trials (Devine et al. 2005), and there are no pharmaceutical industry representatives providing services as "phantom" coauthors. Nonetheless, there is evidence of the same ethical threats and problems documented in biomedical fields occurring in other STEM fields, albeit with somewhat different impacts. Far from being restricted to biomedical fields, problems in scientific collaboration are ubiquitous in science (Shrum et al. 2001; Bozeman and Boardman 2014).

One of the difficulties we face in unraveling the evidence we present here is the difference between ethical problems and just being a jerk. What we shall see is that in each of the chapters, no matter what the category—gender dynamics, work style, or whatever—researchers are called out for bad behavior. In most instances it is easy enough to determine when someone is being a jerk, as compared to being unethical or even illegal. We return to these issues later. We also return to another very important issue. Most of our data provide only one participant's version of a research collaboration. Particularly when ethical issues are at stake, multiple perspectives are useful.

5

Research Collaboration Effectiveness

IN THEIR OWN WORDS

Introduction

There are many disquisitions about the relative status of the "hard" sciences and the social sciences and many of these end up concluding that "social science" is an oxymoron—social yes, science maybe not. However, the social sciences do have one, perhaps only one, clear advantage over the natural and physical sciences: our phenomenon talks back to us in relatively plain language that we sometimes understand. While it is not always that easy to sort out our subjects' subjective responses, we are nonetheless advantaged, and, in case of the present work, it is our subjects' words that provide most of the knowledge presented here.

The Aggregate Model of Research Collaboration Effectiveness presented in the preceding chapter identifies major causal factors. Here we put the model to work, considering each component of the aggregate model against the observations of the practicing researchers who provided data about their own collaborations. They tell their stories; we simply organize them in a way that we feel reveals as much as possible about research collaboration effectiveness.

Disciplinary and Interdisciplinary Dynamics

In considering the determinants of research collaboration, a reasonable working rule is that the more sweeping and external (the collaboration) the more difficult it is to pin down its impacts on collaboration outcomes. If one researcher does not like the others' work habits, that is knowable, and we can rely on the researcher's report to be valid, at least with respect to perceptions. But when trying to trace disciplinary impacts or interdisciplinary factors, the causal path may be winding. For example, if a person in a discipline with relatively few intellectual property stakes (e.g., economics) has trouble working with someone from a field with considerable intellectual property stakes (e.g., electrical engineering), is the dispute really owing to differences in disciplinary traditions, norms, and training or is it owing to the simple fact that one party is worrying about IP (intellectual property) and the other is not? Or is it both at the same time? Similarly, some disciplines rely extensively on large-scale equipment and depend on massive funding of facilities, whereas others do not. Is it the discipline or is it equipment dependence, or is it simply resource needs driving the differences among collaborators?

We know that fields and disciplines do have some independent effects, inasmuch as fields have unique histories, distinctive professional norms, and socialization patterns (Austin 2002; Moody 2004), but it is not easy to sort out those effects. When disciplines are brought together in research collaboration, the complexities multiply with the number of disciplines (Qin et al. 1997; Katz and Martin 1997; Jeffrey 2003). According to the research literature, differences in perceived status and status allocation processes often cause problems, as does disciplinary parochialism (e.g., Becher 1994; Younglove-Webb et al. 1999). Mutual respect for the collaborating disciplines cannot just be assumed.

Cummings and Kiesler (2005) provide some particularly useful research on cross-disciplinary boundaries. They find that coauthors indicate no greater problems in cross-disciplinary collaborations than in single discipline collaborations but that collaborations spanning university boundaries more often have negative outcomes, especially when those multiuniversity collaborations are multidisciplinary. They provide several suggestions for enhancing research collaborations across disciplines and institutional settings, including increased face-to-face interaction, tools to support simultaneous group decision-making, and careful tracking and planning of tasks during the life of the collaboration.

Cummings and Kiesler (2005) contend that many problems pertaining to researchers associated with multiple universities and from different disciplines could be remedied with more coordination. But the solution is not reliance on administrative "higher powers," and, indeed, the authors find that that having the researchers' respective university administrators involved in a project "significantly reduced the likelihood that PIs would actually employ sufficient coordination mechanisms" (Cummings and Kiesler 2005: 715).

One of the trouble spots in collaboration that may be affected by researchers working in different institutions and from different disciplines is coauthoring and crediting. The issues interact sometimes with funding competitiveness among disciplines and institutions (Devine et al. 2005; Drenth 1998). Ultimately, authorship and crediting work best when there is mutual experience and genuine trust, and these may be more difficult to cultivate when collaborators are from different disciplines and with different training and diverging norms (Lagnado 2003).

Turning to our interview data, consider the following response from an individual whose specialty is techniques for gene expression. The work performed by this individual contributes to many fields of biology, but persons working within these other fields seem to have little understanding or appreciation of the difficulties of the work, with the result that our respondent feels that people he helps sometimes take advantage of him and do not share credit:

> Recently, we did a lot of work with another university, brought in a person from their university as a postdoc, backed up the data with [the specific technique], but we got no credit for it. The reason in part is because this is practical work. People know we do this work and they contact us and say "hey, can you just set up a gene expression for this [research project]. It will be easy for you to do. We just want you to confirm our results by using another expression approach." They don't know much about this approach so they just assume it is easy to do even though it isn't. Then they don't give us any credit.

One well-known collaboration issue, related indirectly to differences among fields and disciplines, is the differential cost of research work. As one interviewee, a mathematician working with a chemist, notes:

> Chemistry is very much driven by resources, sometimes more than actual interest in the science. But my work is very autonomous, driven

only by the science, because it is so math oriented and, as you know, math is cheap. We don't need lots of expensive equipment to do mathematics work.

In many cases, the cost issue affecting collaboration is the cost of large-scale equipment. Equipment cost sharing can lead to collaborations that are forced fits. As described by an engineering professor:

It [effective, timely collaboration] is really a mix centering on whether you use a large piece of equipment or not. There is a lot of effort to make us collaborate and sometimes that does not work.

In general, cost issues are not often mentioned in reports of collaboration problems. However, anothersize-related issue mentioned often is the number of collaborators associated with large-scale projects and the profusion of authors. Many note that sorting out credit is a special problem in large projects. An astrophysicist's comments below are illustrative:

Our experiments involve people around the world. That tends to be because sites of experiments are around the world. The size of collaborative groups ranged from a dozen to more than 100 in our NASA-NSF project. In astronomy we try to keep few authors as the number and position of authors are important on a CV. If we can do an observational paper, that is important. In the NASA project we use a random name generator—the big papers vary the first authorship and keep the rest.

Research on interdisciplinary work has shown that while there are often great benefits to bringing together people from diverse disciplines (Holland 2013; Baker 2015), the process sometimes leads to disagreements, due to differing norms and expectations and perceived statuses (Rossini and Porter 1981; Öberg 2009; Garner et al. 2013; Parker 2015).

In our interviews there were relatively few reports of disciplinary clashes leading to bad collaboration outcomes (indeed, many respondents characterized working with scholars from different disciplines as bringing in "complementary capabilities"). Interestingly, almost all such reports incorporated negative experiences working with medical researchers. This is especially telling, inasmuch as our interviews specifically excluded medical research and medical researchers (though in retrospect it would have been interesting to hear "the other side" of the collaboration—the experiences of medical researchers working with engineers, chemists, and biologists). Here is one of many available examples, this one from a biostatistician:

The collaboration I had, what made it bad was working [with physicians] on more of a consulting basis, working with some medical doctors and doing analysis of their data. It went badly. They were very rigid on what they wanted but then they didn't know. They thought they knew what they wanted and they wanted an analysis done but they didn't really understand the questions that needed to be asked. . . . The physicians collected the data somehow through their practice or some money that they had through the hospital and then really had no idea how to do any of the analysis of it or set up the study design. . . . I would end up really having to reorganize the data in a certain way or ask them a lot of questions and then sometimes they get very defensive. . . . I believe that they just recognized me as providing a service rather than an expert at something. It was always just like, "Do this work. We need this done." But they had no idea that what they were doing was not really research. It has kind of been a nightmare dealing with the physicians.

The following response is particularly useful because it relates to a problem that is apparently growing: collaboration difficulties owing to affiliation with an interdisciplinary center and multiple problems associated with multidisciplinary work in a multidisciplinary setting (Suh 1999; Boardman and Corley 2008; Ponomariov and Boardman 2010):

I have been in collaborations where I spent a lot of time discussing work with people but there's nonetheless some misunderstanding and for one reason or another the project never really goes anywhere. . . . I do have one collaboration that has not been at all effective but I'm not even sure it's a collaboration. I am a formal member of a [subfield] Center but I don't really do much of anything with them and my research doesn't go through the Center. They just wanted to have my name on the center. I don't think collaborations dictated from the top are ever very effective. People don't collaborate well just because there's funding available. Good science doesn't come out from those sorts of directed collaborations [involving researchers in different disciplines at different institutions].

Commerce and Interinstitutional Factors

As is the case with most external factors related to collaboration effectiveness, it is difficult to pin down the causal route in the cases of persons from different institutions working with one another. One possibility is that if the collaborating institutions have more in common with academic settings, such as basic research-oriented federal laboratory research, as at Brookhaven National Laboratory or Lawrence Berkeley National Laboratory (see Bozeman and Fellows 1988; Leyden and Link 1999), strife may be less than one would find with vastly different organizational cultures and with quite different missions. In particular, we might expect special hazards in research between university faculty and researchers at industrial firms, given the divergence in institutional missions and in such organizational factors as work autonomy and job security (e.g., Hall et al. 2001; Goldfarb and Henrekson 2003). If interinstitutional collaborations are more perilous, an interesting question is whether the potential for bad outcomes relates to differences in norms and procedures or to less interpersonal familiarity among participants, or to the transactions costs of dealing with larger and more remote collaborations (Garrett-Jones et al. 2010).

Some researchers (Liyanage and Mitchell; Bstieler 2015; Boardman and Bozeman 2015) provide evidence that one of the major factors in bad outcomes from multi-institutional research collaboration is the importance and centrality of commercial concerns (for an overview, see Perkmann et al. 2013). To be sure, it may also be the case that commercially oriented collaborations will be particularly positive because they promote enhanced social capital and doing work outside the routine (Tartari and Breschi 2012). However, when commerce-based problems arise, they often generate some of the very worst collaborations, including Nightmare Collaborations.

Focusing more broadly than on particular disputes about intellectual property, many have noted possible clashes between the norms of commerce and the norms of science (Rai 1999; Rappert et al. 1999; Walsh et al. 2007). Case and historical evidence shows that the clashes can sometimes be highly disruptive (Welsh 2008; Bekkers et al. 2002).

Here are two examples of the concerns that many researchers have with commercial issues, a concern reflected in the research literature (Bruneel et al. 2010; Bodas et al. 2013), concerns about respective time frames and goals of industry and university researchers. One respondent noted:

> I have not been able to collaborate [effectively] with people in industry. They say things like "show me the business value" and that narrows the conversation to the instrumental. When that happens the conversation is over.

We get a different view, but not entirely at odds, from a respondent who spend substantial time in industry before taking a university job as the director of a major center, one, not coincidentally, charged with performing cooperative research with industry. His views are experience based and particularly thoughtful:

> Collaborations in industry are more targeted, and have clear metrics. You work with industry because you want to produce something. Industrial collaborations can be much more effective than in academia, but can be less intellectually engaged than those in academia. Academia is less effective in terms of producing intellectual output. The good thing in industry is you know what you are supposed to do, and there are metrics to evaluate it. In industry, if you do not produce research output, the consequences can be brutal. In academia, it is much fuzzier—and that can become cover for intellectual masturbation. I am amazed how little meaningful thought comes out of academia, given the amount of effort that goes into it. The publication thing is ridiculous—little pieces here, little pieces there. You cannot help but think there is research produced that did not need to get done.

Here is a contrasting view, also from someone with much university-industry experience, but almost all of it from the point of view of someone engaged, though not exclusively, in "the publication thing." Our respondent is a senior biotechnology professor. First, he describes how his research group came to work with a particular industrial firm:

> We had a collaboration with a [foreign] company, but we had lots of problems, including intellectual property disputes. This company was originally a gigantic firm that was trying to diversify and go into other areas including biotech. The way we got involved with them is that they wanted to use a patent that we had developed. They first went to court to try to get the ability use the patent. When that didn't work they figured they would license the patent from us and give us money to essentially work as their biotech lab. Once they started working with us they got the license they wanted in the first place and then they started talking to us about deliverables.

Problems grew when the industrial partner pitted his group against others in tournament-like competition with larger firms and other academic competitors:

> When we gave them the products and the license, we went to [the country] and had conferences and they asked two or three other

[national] groups to come to the conference. We had no idea that these other groups were going to be there. So, we had the conference and then after about six months we found out that they had the [national] groups working on exactly the same things we were working on. . . . The whole thing was only about $200,000, and we were competing with multimillion-dollar labs on the same topic.

At least from the interpretation of the respondent, many of the problems of the collaboration had to do with specific issues related to this specific company (particularism once again raises its head):

I think one of the big problems was that they simply had almost no research and development experience they were just a big multimillion-dollar company trying to become a research company. If they had a better understanding of what they wanted then I think things would've worked out much better. . . . They didn't consider deliverables the same way that we consider deliverables. Sometimes they really didn't even know what they wanted but they wanted us to deliver it anyway. We were supposed to guess what they wanted. After a while things started breaking down and we didn't renew the contract.

Thus, the story above seems more a cautionary tale about problems working with collaborators who are inexperienced in research rather than a general problem working with industry. Nevertheless, inexperience with basic research is much more likely to occur with industrial firms, showing once again the entanglement of causal determinants of research collaboration outcomes.

One in search of good ideas about effective management of university-industry collaborative research can find plentiful advice in the research literature of science and technology policy (see Davenport et al. 1998; Siegel et al. 2003; Sherwood and Covin 2008; Bstieler et al. 2015. For an overview, see Bruneel et al. 2010). The researchers we interviewed provided some excellent advice as well, including the following commentary from an individual who has worked successfully with industry. He suggested that it is easy enough to avoid friction, at least if academics have clear expectations and understand the motives of industry collaborators:

I have gotten some money from private industry. . . . Private industry doesn't publish. They put strings on the collaboration; for example, you have to give them a six months period to review the publication.

You know that private industry's goal is to make money and get as much as possible for free. That is the rule of their game. The problem is when academicians cry because they think the rules are the same for private industry as academia and then private industry doesn't behave that way.

Work-Style Fit

We turn now to the internal (to the collaboration) dynamics that relate to research collaboration effectiveness, most pertaining to team science elements. One of the factors most commonly mentioned when we asked our respondents to tell us about important factors in a successful collaboration is the fit of work styles among the collaborators. Our research shows that a lack of fit of work styles is one of the most commonly mentioned factors in causing bad collaboration outcomes (Bozeman and Corley 2004; Youtie and Bozeman 2014) and, similarly, quality fit is one of the factors most often mentioned when citing successful collaborations and good collaborators (Youtie and Bozeman 2014; Bozeman et al. 2015).

This does not imply, of course, that individuals work best when they work with those who have identical work styles. Just as important is the ability to value others' work styles, even when they are different from one's own. Many of the work style–related problems that occur in research collaborations differ little from the work style problems one finds in most any organization, including ones not in the business of knowledge or technology production (e.g., Kabanoff 1991; Ayoko et al. 2003).

Here is an example where a respondent, a senior female chemical engineering professor, discusses the importance of work-style fit as a key determinant of her collaboration choices:

My rule is that I collaborate with people I like. People who collaborate need to know about what I know about without knowing the same thing. We have to be able to communicate in general terms without actually learning someone else's field. My most successful collaborations have been with [gives name] and [gives name] and those collaborations have been very good indeed. The writing is really the key. If you both take the time to express well and articulate that is a key factor. Also if you like sitting around in a room and bouncing around ideas and doing "what if's" then I like to collaborate with you.

This happy experience contrasts sharply with the "noncollaboration" below, one that died without even getting off the ground.

> Last year I was asked by my department chair to collaborate with a big-name person [at the same university as the respondent]. I said, if he wants to collaborate with me then he should ask me, he should talk to me, and he should send me a copy of the proposal. There is no reason for him to ask you to communicate for him. Finally he [the 'big-name' professor] got back to me and sent me a copy of his ideas for a proposal and it was clear that what he wanted me to do was to work on [a subfield]. I told him that I don't do anything in [this subfield], and that I don't have any interest in working on a proposal that relates to [it]. But he has such a big group in such a big bureaucracy that my response wasn't enough. I still kept getting e-mails from his research staff asking me when I was going to contribute my CV and when I am going to write up this or that. Somehow it has not gotten through to them that I declined to work on the proposal. I wouldn't want to work with this guy under any circumstance. Why are you asking my chair to recruit me for your project, why not talk to me? That's just total disrespect. I don't want to work with some jackass that I don't even know and then work from a distance and communicate only through other people.

One of the most common and pernicious problems in research collaboration is that some people are controlling and some people, probably most people, do not like to be controlled (Fiske 1993; Keltner et al. 2003). Control problems may be especially acute among peers (Freidson and Rhea 1963; Raelin 1989). Many collaborations are undermined by collaboration participants' need to control everything. Here are some examples from our data:

> I was working with a collaborator. He had the data but he did not want to share the data unless all ideas were first channeled through him. I said I couldn't work that way so I dropped it.

Another control-related illustration:

> [Names person] was appointed director of [gives name of NSF Science and Technology Center]. As soon as he was appointed, he had a personality change. His idea of being director was to control everyone that had

anything to do with the center. A lot of us just went back to our faculty homes and quit working there.

As is the case with most of the determinants of research collaboration effectiveness, causality is complicated. Often it is nearly impossible to parse out the causal paths among three related problems: (1) personality issues and personal psychopathologies; (2) personality clashes where there is no real culprit; and (3) problems related to the compatibility of work styles. The following comments from a senior engineering professor seem to us to contain much wisdom:

> I think the most successful people are the ones who are most adaptable, who can say, okay I have talent here, but I need to adjust my expectations. I have a student who is not a morning person but who is brilliant. He just cannot get in before ten A.M. When he has to show up for an early morning meeting, you can see it in his eyes that he is not awake yet. I think the real skill in being productive when you are working with people you have to ask, "What drives this person? What are they bringing to the table? What's their baggage? How do I adjust my expectations?"

Career Stage

One's career stage relates to collaboration effectiveness, not least because people at different career stages often have different needs, and these needs may clash. We can also consider whether persons who are earlier in their career are more likely to be exploited or, since they often are less secure, are more likely to be the cause of collaboration disputes. If researchers experience bad collaboration experiences early in their careers, is it because they are more likely to be exploited, because they are more likely to make bad collaboration choices, or because they have not yet had time to develop collaboration and interpersonal communication skills?

One complication of career stage issues and their effects on research collaboration effectiveness is that they are often intertwined with age issues. As one might expect given the convenience of measuring age, there are some relevant research findings. However, the studies focusing on the relation of age to collaboration show mixed results. For example, Ponomariov and Boardman (2010), in examining academic faculty affiliated with university research centers, find no significant relationship between researchers' career age and their number of publications with industrial collaborators, at least not after controlling for a number of potentially confounding variables. This is perhaps not so surprising as it appears at first blush. The percentage of faculty, young and old, publishing with industry-based researchers is relatively small, 11.4 percent for those not affiliated with centers, 20.7 percent for those affiliated. A greater complication: those working with centers, even younger researchers, are sometimes brought into the centers *because* they have a history of working with industry.

Bercovitz and Feldman (2008) present some unexpected results in their study of the commercial activities of medical school faculty. The likelihood of being involved with patenting and licensing *decreases* with age, at least for this group. Haeussler and Colyvas (2011) report a set of more common findings. In their study of life scientists, they found that older scientists are more likely to be engaged in a variety of commercial activities, including not only patenting and licensing but also consulting and founding a firm. In another S&T Human Capital study, Lee and Bozeman (2005), examining more than six hundred academic scientists in the United States, find that career age interacts in such a way to mitigate the relationship between collaboration and productivity. Younger and mid-career researchers are more productive in terms of publications per collaboration, but once a certain age threshold is reached the older researchers begin to have fewer publications per collaboration. While Lee and Bozeman do not investigate this hypothesis, it is possible that older researchers have more mentoring-based research collaborations with students and postdocs, collaborations that may have greater payoff in terms of human capital development but less in terms of publications per collaboration.

Aschhoff and Grimpe (2011) look at age effects in terms of possible "imprinting" effects of young researchers who work with industry. They examine citation data for 343 German academic scientists in biotechnology and find that if young researchers have coauthors who have publications with industry personnel, then they in turn are more likely subsequently to have publications with industry. They conclude from this evidence for strong peer effects, but, of course, it is also possible that these are strong selection effects, positively choosing senior coauthors because of the commercial work or industry contacts.

The effect of industrial work on collaboration patterns is one of the most common topics in studies of age and career impacts on collaboration. But in actuality the percentage of researchers, especially young researchers, working with industry remains small. Most of the interview data we obtained related to a wide variety of career issues, but most common were issues related to tenure, promotion, and reputational goals. Let us first consider the reflections of a senior faculty member concerning her changes from early career collaboration criteria to those later in her career:

> When I was an assistant professor I was much more instrumental and focused on what a person could do for me than whether we could work together easily. Now I am less interested in who can get me ahead or even who can do great science: I want to collaborate with people I like.

A tenured computer science professor gives a similar perspective on the course of careers and effects on collaboration:

> When I was younger, I focused more on substance and the needs of a particular project. I thought about how people could help me do a project, make up for my own weaknesses. I was interested in people being complementary to me. It was much more instrumental than it is now. Later in my career, I am moving away from instrumental and more toward enjoyable collaborations. I think this shift is true in general later in one's career: you do things that are enjoyable. Now, what I look for in collaborations: Is the person worth working with? How much will I learn from this collaboration?

Of particular interest is the role of tenure seeking in bad collaborations (Hearn and Anderson 2002; Thursby et al. 2007). In some cases individuals are under extreme pressure (real or perceived), and this pressure could have the effect of promoting bad behavior as well as shaping perceptions about others' bad behavior in collaborations (Floyd et al. 1994).

Most untenured professors at least feel vulnerable and in many cases are, indeed, at real risk of losing their jobs. In some cases this can lead to bad behavior, partly from desperation. As is often the case in human relations, the one suffering from the bad behavior of a vulnerable person is one who is even more vulnerable, in the case below, a student. In a previous chapter we cited the case of a professor seeking to hold on to a student researcher, against his will and despite the fact that he wished to leave his doctoral studies, solely because of the self-interests associated with her impending tenure decision.

Whether or not strategies and motives change during the life course, it seems likely that bad outcomes are especially likely in early career collaborations:

When I was first starting out as an assistant professor about half of my collaborations were bad ones. I was very flattered to work with people who are well-known but then I would get into the collaboration and think oh my God what have I done? But I haven't been in a really awful collaboration in a very long time.

Often, students are among the most vulnerable, as related by one of our senior respondents:

[I know of] a student who worked in a colleague's lab a few years ago. He really wanted to get a master's and she wanted him to get a Ph.D. She talked him into getting a Ph.D. In his third or fourth year he said, 'I don't want to do this anymore. I want to leave.' It started really going downhill from there. She was coming up for tenure and needed to have both a certain number of students as well as publications. She was pushing him to finish multiple papers at that point. She was refusing to allow him to graduate until he finished the papers.

This last anecdote makes us wonder where the other professors were in advocating for the student. In general, we found that collaboration dynamics that were related to rank differences had to be sorted out by the collaborators themselves, and often by the most junior collaborator. These experiences do lead to the development of career-stage heuristics, but it does beg the question of whether people leave the field as junior researchers because of power dynamics related to career stage.

Trust and Experience

Previous research (e.g., Melin 2000; Bozeman and Corley 2004) tells us that researchers often have strong preferences for focusing most of their collaborative work on those with whom they have had previous collaborative successes rather than developing the broadest possible collaborative network. This may be especially the case with female researchers (Fox and Ferri 1992; Corley and Gaughan 2005). In some cases, collaborators, through long working experience, have such a close reading of one another's values that direct and explicit procedures may be unimportant. Relatedly, most researchers feel that interpersonal trust is a strong factor in effective collaboration (Melin 2000; Davenport et al. 1998; Bennett and Gadlin 2012). The role of experience in collaborations is closely intertwined with trust, and usually the relationship is reciprocal. Collaborators have more trust because they have experience with one another, but, at the same time, if the trust is sustained it likely enhances the propensity to continue collaborating.

When we asked researchers to tell us about their best collaborations, trust was a common feature. Thus, one respondent notes:

> I have one colleague I have worked with since we met at a conference in the early 1990's. We hit it off, had a lot of the same interests and I found it easy to work with him. The more I worked with him the more that I saw that I could count on him. It was not just that he would do his part but I could count on him. I could be sure he would do quality work, he would worry about the students we might have on the project, he would be patient when we ran into obstacles. I just trust him. Even though I have had a bunch of collaborators, I still work with [gives name] every time I get a chance.

But trust is not always about happy outcomes. Consider this respondent's commentary on "lost trust":

> [The worst collaborations are the ones with] lost trust. I can think of two examples, in one everything I write is thrown back at me, I get an email about everything. The other example is when I lost faith in a person I should have trusted. After I step away from the collaboration I saw I was the problem.

As is so often the case, the trust factor blends with others. Thus, people who have the same disciplinary norms may find it easier to trust one another compared to collaborators who have different disciplinary norms. In other cases, the importance of trust relates to the conditions under which the collaboration is structured. The passage below suggests as much:

> There are different types of collaboration in the field [biology]. You have a joint grant where you are required to work together. There are other collaborations that are less structured, so if those aren't going well, people can just leave. A shared grant means people have to work it out and people have to trust each other more. So they lead to different conditions in which you suggest to collaborate with one another.

Gender Dynamics

Gender issues play out in remarkably different and complicated ways in society, in the workplace, and in research collaboration. Gender is often intertwined with many of the other factors related to research collaboration and its effectiveness. For example, women academics tend on average to be younger because women have only recently entered the academic science careers in large numbers. Thus, women team members often are at different career stages, most frequently playing a junior role in teams led by older persons, usually men. As we have shown elsewhere (Bozeman and Gaughan 2011; Gaughan and Bozeman 2016), it is sometimes exceedingly difficult to sort out direct effects owing to gender from effects mitigated by or spuriously related to gender. However, our previous research and much of the research literature (e.g., Kyvik and Teigen 1996; Fox and Stephan 2001; Fox 2005; Abramo et al. 2013) on collaboration give sufficient evidence (e.g., that we feel that gender often is a distinctive feature governing collaboration outcomes.

Much of the literature on women's STEM careers recounts obstacles and tells troubling stories, and most, not all, of the gender-related data we obtained from interviews and Web posts were negative. Studies (Long and Fox 1995; Long 2001) show that women progress more slowly through academic ranks and that they are more likely to be denied tenure. As in most walks of life, women researchers are on average paid less than men academic researchers, even when taking into account the fact that women overall have a lower median age and career age. Women have fewer opportunities to participate in industrial collaborations and multidisciplinary research centers, despite evidence that doing so is generally beneficial to careers and advancement (Corley and Gaughan 2005).

In some cases these issues reported by our respondents play out exactly as reported in the feminist literature on occupations and workplace dynamics, with nothing specific to STEM careers; in other cases the issues were more distinctive to academic science. It is easy enough to identify gender issues but exceedingly difficult to sort out causality. There is a particular acute possibility of attribution errors when we are dealing with a fixed characteristic (here gender) that is invariant and has broad effects. That is, it seems highly likely that in some cases people attribute problems to gender when they are a lesser causal factor than other personal or situational attributes (Hewstone 1990; Beyer and Bowden 1997; Hoyt and Burnette 2013). Is the problem because I am a woman or am I a woman who is experiencing a problem, but not because I am a woman? Nevertheless, there are many gender-related collaboration problems that relate not at all to attribution errors and misunderstandings but, rather, to either the gender dynamics among individuals (Probert 2005; Lester 2008) or—perhaps even more worrisome—the failures of institutions and organizations to guard against discrimination and potentially preventable bad behaviors (Bailyn 2003; Eveline 2004).

Not all gender-related stories are negative. Consider the respondent, quoted below, who recounted the negative feedback she received from some of the faculty members in her department who assumed she would not complete her PhD when she had her first child:

> I had two children when I was in graduate school, and so I had two maternity leaves. Science is very competitive, and the project I was working on was very competitive. There were others outside our group working on it, so it is not unheard of for an investigator—if someone is on maternity leave—to take that project and give it to someone else. And then if you have done all the legwork to get it working, you lose credit for that in the publication process. To her credit, my advisor did not give my project away, despite the fact that she was then going up for tenure [and needed quick results]. I felt that she was really being supportive, that she was willing to wait, that she was willing for me to do a decent maternity leave. I was literally gone for three months the first time and six months the second time—I had some complications. I came back and my notebooks were there.

The above story at least seems like a happy one, and at one level it is, but it also underscores two problems. First, it seems, at least from the evidence provided, that one's treatment in such a case is not a matter of routinized policy but the "luck of the draw" in supervisors. Our respondent had a lucky draw. Second, the case shows one of the reasons why people who are having babies (or who are facing similar life transitions) sometimes have unfortunate collaboration outcomes. If junior faculty or students are working for faculty whose very careers may depend on the result produced by the person being supervised, not everyone will be self-sacrificing. This implies, perhaps, that such situations beg for institutional solutions, not individual ones.

The expectations problem is not confined to individual collaboration dynamics or even to particular university settings. Apart from any particular university, careers can suffer when productivity is affected by family responsibilities. The same woman who experienced such support from her own dissertation director goes on to explain that her family-related decisions nevertheless have had lasting career consequences:

> NIH has rules [pertaining to consideration of] maternity leave. But the review panels are not going to give you a break. My child was born at the very end of my graduate training, and then I was writing my dissertation while I was on maternity leave, defended, finished up some papers. But I had a gap in that year in my publications. My reviewers when I submit for grants still complain about that.

In some cases there are institutional and public policy protections already in place, and they are still not sufficient to the task. Moreover, some of the most troublesome gender problems encountered in the academic workplace have much in common with those found in any work setting. One such problem is sexual harassment, a focus of much attention by policy makers but a problem than has not been eradicated. Obviously, effective collaboration requires some level of confidence in one's personal security. Consider the "stalker mentor" case summarized in chapter 2, which rises to the level of a Nightmare Collaboration, in our assessment, about the postdoc whose direct supervisor fell in love with her and stalked her and who, when she reported it, ultimately had to leave the research group and was stripped from her in-process publications. Note that we do not consider this a Nightmare Collaboration because of the sexual harassment—although surely that was not pleasant. It is a Nightmare Collaboration because the respondent's attempt to address the sexual harassment problem resulted in punitive action that had a direct impact on her academic research productivity and reputation. This led to a series of unfortunate institutional responses that led to significantly altering the trajectory of this scientists' career, with the researcher having to take a lesser, nontenure track position that offered less job security and respect.

Sometimes discrimination-related collaboration problems interact with one another. Nevertheless, some people have managed successful careers and harmonious collaborations despite dual minority status. A senior female researcher, who is also a member of a minority group, noted that it is possible to manage others' biases and expectations but that it is not easy to do so:

> There are few females in life sciences that are in [indicates her own highly specialized area]. So people aren't accustomed to working with women in my subfield. I have encountered many gender issues, not so many

race issues, which is good since I could get the double whammy. A lot of it is this is sort of a 'respect your elders' thing and you could see the differences in expectations between young males and young females. It is okay to be an aggressive young male but not an aggressive young female. A young female can't just say to a senior scientist 'I think you are wrong.' Even if you say it in a very nice way you can still be sure that everything just goes to hell. I did find when I was a young scientist that I could work well with men, even very senior and aggressive men, but only if they perceived me as a sort of daughter and not as a competitor. Maybe it helps if they have daughters themselves. I don't know.

Perhaps related to different competition dynamics, one of the most common collaboration problems identified by women pertains to credit sharing, and not just for coauthorship. A senior female researcher reported:

> Women generally don't get credit for what they do. They can make teams a lot better, but that doesn't mean they get credit. There are always problems in open meetings where women have an idea, nobody responds to the idea, and then the man says pretty much the same thing and everybody thinks it's his idea and wonders why women don't make a contribution.

As demonstrated in this section, there are specific collaboration problems that women experience in addition to the ones that both men *and* women experience. These include gender dynamics related to childbearing and sexual harassment. To be sure, men are sexually harassed, but the experience is much less common. We found not a single case of a man suggesting a sexual harassment problem related to collaboration.

Yet despite the importance of gender dynamics, in reflecting on all of our data we found that the men and women share more in collaboration dynamics, both the good and the bad, than they experience differently due to specific gender dynamics. Indeed, one of the primary-seeming gender issues may in part be masked by an understandable but difficult to measure attribution error. Given the structure of academic work today, men tend to be more often senior and more often more powerful. Certainly gender may interact with power dynamics. But the age gap and perhaps the power gap between the genders seem to be narrowing. As that narrowing continues, perhaps negative gender dynamics in collaboration, and the workplace generally, will diminish.

Findings and the Aggregate Collaboration Effectiveness Model

As we see from the foregoing, each of the components of the Aggregate Collaboration Model we presented in the previous chapter not only has connection to the literatures on research collaboration and team science but is also identified (without our specific prompting) by several of the researchers we interviewed. However, it is also the case that the various aspects of the model interact in complicated ways, ones that cannot fully be sorted out with the data we have on hand.

The Aggregate Model proposes that collaboration management and decision-making are almost always important in mitigating other features of collaboration, both internal and external ones, and this view is reflected implicitly and sometimes explicitly in respondents' observations about collaboration effectiveness. The next chapter explores some of the empirical dimensions involved in the management of collaborative research teams, and the final chapter offers some observations about management approaches, as well as some prescriptions. Later, in our concluding chapter, we focus again on issues related to collaboration management, relating particularly collaboration management approaches to collaboration effectiveness.

6

"Decision Making in Collaborative Research Teams"

Introduction

Research collaborations are not tangible things but rather social interactions among real human beings with complicated motives. Nearly everyone who studies research collaboration or has much experience in collaborating understands that institutions and policies external to the collaborating individuals often have important effects, but that when all is said and done it is the individual researchers who make or break most collaborations. Indeed, several studies find that the best collaborations emerge organically between individuals and not from being "forced" by institutions (Chompalov et al. 2002; Melin 2000).

As we noted in previous chapters, one of the most common approaches to studying research collaboration, especially individual-level collaboration, is through the lens of bibliometric analysis of formal coauthorship patterns (Melin and Persson 1996). However, such studies, even if they provide indirect evidence relevant to research collaboration effectiveness, typically provide no direct evidence. Nor do bibliometrics studies accommodate the truism that there is more to research collaborations than just the coauthorships published in a journal article (Katz and Martin 1997; Sonnenwald 2007).[1]

In this chapter we examine factors related to decision-making in and about research collaboration, not for simple idle curiosity but rather to determine their relationship to collaboration effectiveness. We explore these motivations and processes in two ways. First, we ask respondents about their most recent coauthored paper, an approach that has the advantage that those events are easily recalled.[2] Second, we ask respondents to tell us about their career-long experiences with collaboration. The chief advantage here is to obtain a broader perspective and to allow us to ascertain the likely probability of certain collaboration-related experiences, especially ones that are not commonplace. We begin with an analysis of pre-collaboration relationships: How do individuals choose collaborators? What are their motivations for doing so?

With Whom Do I Collaborate? Pre-collaboration Relationships and Motivations

The literature on pre-collaboration relationships and motivations draws on group socialization studies. For example, Levine and Moreland (1994) find that new group members feel excluded unless existing team members expressly welcome them. In contrast are studies that examine the actual initial event that led to the collaboration. Goel and Grimpe (2011), for example, distinguish between active and passive networking among German academic scientists. The authors operationalize active networking as participation in academic conferences, while passive tends toward collaboration with colleagues from graduate school or well-known current colleagues.

Many of these studies focus on the effects of demographic characteristics, such as career stage or gender. In the case of career stage (which itself is intermingled with physical age and cohort effects), Lee and Bozeman (2005) find the career age mediates the relationship between collaboration and productivity. However, Bozeman and Corley (2004) report that tenure does not weigh heavily in collaboration choice because of the mentorship relationship of senior scholars with junior scholars or students. In the case of gender, Bozeman and Gaughan (2011) observe, from a survey of 1,700 respondents weighted by field, that men are oriented toward collaboration based on instrumentality and previous experiences.

Studies of the intangible characteristics of collaborators and their interpersonal relationships are less common. Collaboration motives are difficult to distinguish because academic researchers juggle multiple pressures that can be linked back to multiple motives (Senécal et al. 2003; Boardman and Bozeman 2007; Kyvik 2013). We examine motives using similar questions to those asked in Bozeman and Corley (2004), Lee and Bozeman (2005), and Bozeman and Gaughan (2011). These questions give rise to typologies of research collaboration motivations based on productivity, timeliness, similarity of language, career stage, friendship, and ability to make a distinctive and complementary contribution.

MEET AND GREET

One obvious issue of importance to collaboration is the "meet and greet": where did individuals meet their collaborators? The most common (see figure 6.1) way was through the student-professor relationship. Nearly 40 percent of respondents met in this way, where the coauthor was a student, former student, or postdoc. The second most common way, by more than 30 percent of respondents, was meeting through being in the same academic department. The third most common way was meeting at a conference, with more than one in four respondents having met this way. Least common was having met in school (which was indicated by only 13 percent of respondents). These findings suggest the importance of geographic proximity, which may seem obvious but is still worth emphasizing in this age of information technology. Perhaps future generations will rely less on interpersonal networks to form relationships for research collaboration, but face-to-face connections continue to remain important today.

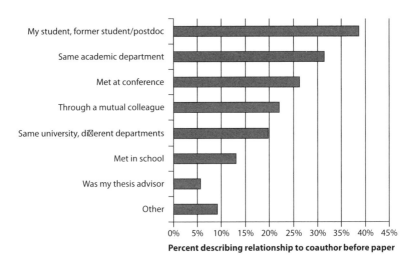

FIGURE 6.1. "How would you describe your relationship to your coauthor(s) before the collaborative paper?" (Percentages exceed 100 percent because of multiple types of relationships)

TO WHAT END? MOTIVATIONS FOR COLLABORATION

Once researchers meet, what leads them to decide to collaborate? Concerning collaborating on the most recent coauthored research paper, we developed a list of possible motivations for collaborating.[3] Unlike with many scales, respondents tended to have strong opinions on their motivations for collaboration in that there was notable use of the extremes on the scale. This list was similar to that used in Bozeman and Corley (2004), Lee and Bozeman (2005), and Bozeman and Gaughan (2011).

What was the most important factor in collaborations? The answer is "Working with researchers whose skills and knowledge complement mine." More than half of respondents rated this 10 or "extremely important" and another 20 percent of respondents rated it a 9. The next most common factor was increasing my own research productivity; 34 percent of respondents rated this factor a 10 and another 22 percent rated it a 9. These results conform closely to the responses we received from our interview data.

The next two factors are less clear in their relative ranking. These are "working with persons I can depend on to complete work on time" and "having fun working with researchers I like on a personal basis." Completing work on time has a higher mean (7.3 versus 6.9 for having fun) and fewer respondents rating it as 1, or not important at all (5 percent versus 10 percent for having fun). However, having fun attracted slightly more respondents rating it extremely important (25 percent versus 22 percent for completing work on time). These two factors represent the yin and yang of research—working and having fun doing it. Whereas Bozeman and Corley (2004) report a strong gender difference on the criterion choosing coauthors because it is fun to work with them, with men being much more likely to provide that response, our current data show no significant difference by gender.

The next two factors were providing "research training for coauthoring students or postdocs" and "working with collaborators whom I felt would be fair in coauthor crediting or order of authorship." Both of these factors had similar mean ratings: 6.4 for student research training and 6.3 for fair crediting. Of the respondents, 30 percent rated student research training to be the extremely important, which was the third highest share of respondents giving a factor a rating of 10. This result further emphasizes the vital role of student research training to these faculty respondents. At the same time, fair crediting, which only attracted 17 percent of extremely important ratings (i.e., ratings of 10), also attracted another 32 percent of respondents, who gave this factor a rating of 8 or 9.

TABLE 6.1. Motives for Engaging in Collaboration

Motivations for collaborating	Percentage of Respondents Giving Rating:										Means	Observations
	1 "Not important at all"	2	3	4	5	6	7	8	9	10 "Extremely important"		
Helping me with Tenure and Promotion	53.8	2.4	3.1	4.7	12.0	3.3	3.1	6.3	3.1	8.1	3.5	534
Helping coauthor career	21.2	2.8	4.1	3.6	21.9	2.4	6.1	10.1	9.9	17.9	5.7	612
Language fluency	74.1	5.8	4.9	1.3	6.5	1.9	1.1	0.9	1.5	2.1	2.0	547
Training students or postdocs	20.5	1.3	2.4	1.6	14.7	3.9	6.0	7.5	12.5	29.6	6.4	602
Fun working with person I like	10.0	1.7	3.5	2.2	18.1	2.6	9.3	12.5	15.2	25.0	6.9	621
Increasing my own work productivity	3.2	1.7	2.0	1.8	12.7	1.9	8.8	12.6	21.8	33.5	7.9	630
Complementary skills	2.1	0.5	0.7	0.8	6.0	0.7	4.7	11.5	20.4	52.7	8.8	627
Persons complete work on time	5.1	2.1	1.4	1.8	20.5	3.7	6.7	15.9	21.3	21.5	7.3	623
Fair crediting	10.1	5.5	6.3	2.3	22.6	2.5	5.0	13.1	15.0	17.4	6.3	611

Question: "Researchers may have many different motivations for collaborating.... Regarding the factors that led you to collaborate on this most recent coauthored research paper, how important, if at all, were the following?"

The least important factors, according to respondent ratings, were helping a coauthor's career (mean of 5.7), "helping me to obtain tenure or promotion" (mean of 3.5), and "working with persons highly fluent in my native language" (mean of 2.0). Indeed, the latter two factors were comprised primarily of respondents who rated the factor a 1, or "not important at all." More than half of the respondents assigned this rating to the tenure and promotion factor; unsurprisingly, this finding is associated with the respondent's rank in that 76 percent of full professors gave it a rating of 1. Similarly, more than 70 percent assigned this lowest rating to lack of language fluency. The longstanding internationalization of science, both in the ongoing graduate students from outside the United States—accounting in 2014 for 33 percent of all graduate students and 38 percent of all postdocs in science and engineering according to the National Science Foundation[4]— and (2) in international research collaboration—which accounted for 13 percent of scholarly articles in the Science and Social Science Citation Index of the Web of Science in 2000 and grew to 19 percent of these articles by 2013[5]—is certainly behind this lack of concern about language fluency.

Coauthorship Decision-making Process

The literature suggests three coauthorship decision-making paths. The first concerns explicit management of decisions; the second follows disciplinary norms; and the third organically evolves without an explicit process. Chompalov and colleagues (2002) define four leadership styles: bureaucratic, leaderless, nonspecialized, and participatory. The first two are more common in large multi-institutional collaborations, while nonspecialized are associated with multidisciplinary collaborations. However, the authors conclude that hierarchy is not mandatory and participation is usually accepted. Cummings and Kiesler (2005) find that physical meetings and other coordination mechanisms increase the productivity of collaborations. In a similar vein, Duque and colleagues (2005) observe that collaboration decisions are not helped by more information technologies. In contrast to these formal mechanisms, Garrett-Jones and colleagues (2010) highlight the importance of trust in their studies of Australian Cooperative Research Centers. Researchers have some level of independence in deciding collaboration partners and conferring of authorship status (Heffner 1981).

There have been many bibliometric studies examining discipline-based coauthorship norms. Many of these studies have been focused on trends in presenting authors in alphabetical order according to their last name rather than by contribution. Waltman (2012) shows an overall decline in the use of alphabetical order taking all fields of science together, with mathematics, economics, and physics making the greatest use of alphabetical methods.[6] Frandsen and Nicolaisen (2010) note a rise in the use of alphabetical order in economics and frequent use of this method in physics, but not in information science. Costas and Bordons (2011) observe the use of the first author position for early career researchers who made a substantial contribution to the research, and the last author position for senior researchers in such fields as biology, materials science, and natural resources. What this tells us is the importance of taking disciplinary differences into account in authorship crediting.

While research has been conducted on coauthorship decision-making, relatively little research focuses on the interrelation of management, collaboration, and outcomes (Vasileiadou 2012). In our survey questionnaire we asked respondents to tell us the primary basis for the ordering of coauthors (fig. 6.2). More than half of respondents said that the primary basis for the order of coauthors reflects one or more of the coauthors' assessment of the importance of each coauthor's contribution to the research. The next most common response—coauthor order is alphabetic—accounts for only 18 percent of responses. Placement of senior authors after junior authors was indicated in 14 percent of responses.

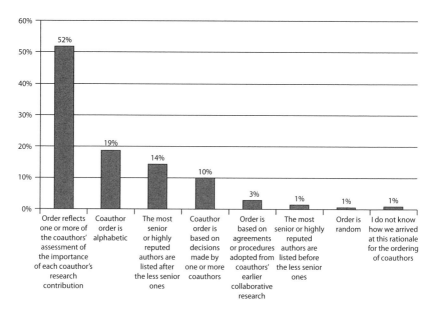

FIGURE 6.2. "What was the primary basis for the ordering of coauthors?" (Number of observations = 637)

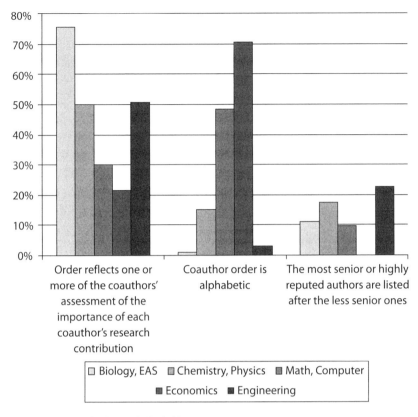

FIGURE 6.3. Basis of author order by field.

As expected, these decisions are associated with a given field (fig. 6.3). Economics and math and computer science are more likely to present alphabetical author ordering; biology and earth and atmospheric science are more likely to present ordering based on author contribution; and engineering is more likely to be ordering on the relationship between junior and senior authors.

One might well ask, "Just why do practices of author ordering matter, since people in different disciplines have standard practices known to those in the discipline?" In the first place, team research is increasingly multidisciplinary research, and these practices can clash if not discussed and understood by all team members. Decisions about tenure and promotion almost always include, at least at some point in the process, people in disciplines different from the person being reviewed. If there is no understanding of disciplinary differences, then people who apply their own discipline's norms may be making unfair assumptions. Finally, it seems likely (though we have

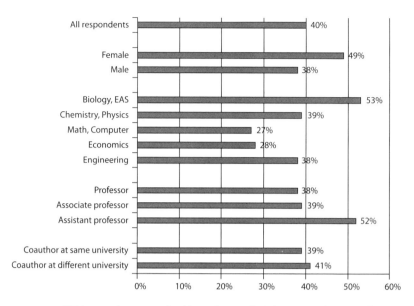

FIGURE 6.4. "Did you and your coauthor(s) ever have explicit discussions about coauthoring credit?" (Number of observations = 641)

no direct evidence) that author-ordering norms may affect the crediting and positioning of honorary authors and knowledge of ordering norms could prove useful in identifying honorary authors.

We move next to results concerning the decision-making path that embraces explicit decision-making and the collaborative team's hierarchy. In our questionnaire, we asked whether the members of the collaborative team had at any point during the collaboration explicit discussions about crediting, such as who should be listed as a coauthor and in what order. Of recent papers, 40 percent had an explicit discussion about coauthoring credit (fig. 6.4). We thought that there might be a difference in having explicit discussions by gender, and indeed women were more likely than men to report having had an explicit discussion about coauthoring credit. Nearly half (49 percent) of female academicians reported having such a discussion versus only 38 percent of the male academicians. Likewise, assistant professors were more apt to have explicit discussions than were associate or full professors, and those in biology and earth and atmospheric science were more apt to have such a discussion than were those in other fields. Since women in the sample tend to be younger in both age and career years, the implication seems to be that younger people have higher career stakes and, thus, are more likely to encourage explicit discussion of crediting. What may be a casual decision for a senior, tenured professor often is less so for those still in career jeopardy.

For the 40 percent of survey respondents who had explicit discussions about crediting, we wondered when the discussions occurred. That is, were these discussions held before the research started, during the course of the research, or after the research had concluded? The most common period for these discussions: during the research. Two-thirds of those having had an explicit crediting discussion did so during the research. The next most common period was after the research concluded; half of respondents had an explicit crediting discussion at this later period. Only 31 percent of respondents told us they had an explicit crediting discussion before or at the start of the commencement of the research. We find this quite interesting and believe that the avoidance of explicit discussion before onset of the research has multiple reasons. In some cases, people are inexperienced and not focused sufficiently on just what they are getting themselves into. In other cases, there is little discussion because people have worked together before, often on many occasions, and have developed implicit norms for crediting. In still other cases, teams are headed by experienced researchers who may feel they know best and who do not afford much opportunity for others' input.

Among the 60 percent of respondents who did not have explicit discussions, following the practice from previous collaborations was more than twice as common (31 percent) than having one person making the decision (14 percent). The role of trust, organic collaborations, and path dependency looms large in this finding. Indeed, there is much agreement that trust is a key ingredient in the success of research collaborations.

In our view, the most important finding about coauthor discussion is that in 60 percent of the cases there is no explicit discussion of crediting, a fact that seems to explain at least some of the collaboration problems reported by the researchers we interviewed. In the next chapter we examine in detail the predominant approaches to managing research team collaboration, including the Consultative Research Collaboration approach that mandates such explicit discussions and, we feel, enhances the effectiveness of collaborations.

Research Collaboration Activities

Science has become increasingly specialized, with division of labor emerging as a dominant paradigm (Walsh and Lee 2015). Beaver (2001) notes the emergence of a common model of research collaboration in which the group is led by a principal investigator, usually but not always a senior researcher. The drawback to this structure is that the principal investigator all too often becomes so enmeshed in grant administration as to lose touch with the actual research (Bozeman and Gaughan 2007).

This typical hierarchical division of labor can also lead to problems occurring between senior researchers (who usually provide the funding and initial idea) and junior researchers (who usually work more hours implementing the research idea, through data collection and analysis). Haeussler and Sauermann (2013) show that authorship involves multiple types of contributions, including conceptualization, laboratory work, and provision of materials and data; these contributions are also affected by prior authorship status. Gaughan and Bozeman (2016) provide evidence not only of differential tasks but that the number of tasks and task roles vary by gender, with women often taking on a wider array of research-related tasks.

Against this backdrop, we asked respondents to tell us, for their most recent coauthored publication, whether coauthors were engaged in various research activities. These activities ranged from initially developing the research question to providing or collecting data, to conducting data analysis or testing, to writing part of the text, to reviewing literature, providing grant or contract funding, or administering the laboratory. Knowledge of specific activities is important for a variety of reasons, including relationship of work role to credit received, but also because such knowledge provides insight into characteristic approaches to managing collaborative work.

Since we are interested in the honorary author phenomenon, we included the category "made no contribution to the research even though listed as coauthor." We feel our data provides useful insight into the pervasiveness of honorary authorship. Of the respondents, 3 percent gave this "no contribution" answer. In other studies the estimated percentage of honorary authors is higher, but that may be because many such studies have been in biomedical fields (e.g., Kovacs 2013; Eisenberg et al. 2013), where the practice seems more common. Is 3 percent a small number? We think not. In the first place, the collaborations being reported are not special in any way, they are just the most recent ones, implying that 3 percent is probably a good overall estimate of the likelihood of honorary authors. Second, given the number of research collaborations that occur and the stakes involved,

3 percent may be a worrisome number. In some cases, others may not be getting due credit. Honorary authors may be credited with knowledge they do not have and, worse, may act on it (in consulting or even congressional testimony). But the real issue, not revealed in these data, is how and why the honorary authorship occurs. If it is a tribute to leadership on companion projects, that is one thing, but if it is the residue of intimidation, that is quite another.

Back at the question of roles played in the work, we find that the credit authors receive is to some extent associated with the particular research tasks that they perform. Lead authors (usually but not always the "highest credited" authors) are most apt to have initially developed the research question; we found that 42 percent of these initially developed the research question. Accompanying coauthors tend to have relatively higher percentages, compared to that of lead authors, of providing/collecting data (39 percent of coauthors), data analysis (40 percent of coauthors), writing text (41 percent of coauthors), and literature review (41 percent of coauthors). Provision of data was almost equally common among coauthors and non-authors: 36 percent for non-authors versus 39 percent for coauthors. Data analysis or testing also was an activity almost as prevalent among non-authors (35 percent) as among coauthors (40 percent). Activities of non-authors were most prominent in the laboratory or center administration category. More than half of persons considered collaborators, but not listed as a coauthor, were administering the laboratory or center.

TABLE 6.2. Distribution of Research Collaboration Tasks

Task	Lead author	Coauthor	Not listed as a coauthor
Research question	42.2%	26.6%	31.2%
Data	24.7%	39.0%	36.3%
Data analysis	25.2%	40.2%	34.6%
Writing text	27.6%	41.1%	31.3%
Literature review	26.8%	41.0%	32.2%
Grant/contract funds	29.0%	31.7%	39.3%
Administration	22.4%	21.7%	55.9%

"Please indicate whether coauthors were engaged in the respective activities." (number of observations = 641)

Assessments of Coauthoring Experience from the Most Recent Paper

The biomedical sciences have been the subject of much research about the dark side of collaborations (Rennie et al. 2000; Wainwright et al. 2006; Cohen et al. 2004). As indicated in previous chapters, honorary authors, ghost authors, and duplicate publications are among the authorship-related problems (Levsky et al. 2007). Biagioli (1998) questions whether we should discard current notions of authorship in light of these problems, although he acknowledges the lack of alternatives to it. Little of the work examining problematic coauthoring occurs outside of the biomedical sciences (as an exception, see Shrum et al. 2001, 2007). While problems in these fields loom large—for example, researchers at pharmaceutical firms serving as ghost authors—there are similarly significant problems outside the biomedical field. Lee and Bozeman (2005) find that the more satisfied a researcher is, the more likely the researcher is to collaborate. Conversely, past dissatisfactions may well sour future collaborations, which could cause problems in career entry in STEM research. In sum, few studies focus on problematic collaborations outside the biosciences, and it is important to remedy this gap in research and go beyond case-based or anecdotal reports of problematic coauthoring and crediting.

Our survey questionnaire investigated the extent of problematic collaborations by asking respondents about their coauthoring experience in their most recent paper. Respondents were asked to indicate their level of agreement with a series of statements about possible outcomes. The list of statements began with three motivations for the coauthorship: scientific, commercial, and societal. For the scientific motivation, the survey asked respondents to indicate their level of agreement with this statement: "I consider this publication to be one of my best in terms of scientific quality." Commercial and social problems were represented through statements asking about the level of agreement with one or more coauthors being strongly motivated by commercial possibilities and about the potential of the research to help solve social problems. We also included a statement to gauge the level of contribution of the respondent to the recent paper.

The crux of this question was centered on three statements about contributorship to the most recent paper. The first was whether there was at least one person who deserved coauthor credit but did not receive it. The second was whether there was at least one person who did not deserve coauthor credit but received it. And the third was whether there was significant gender-based conflict among the coauthors.[7]

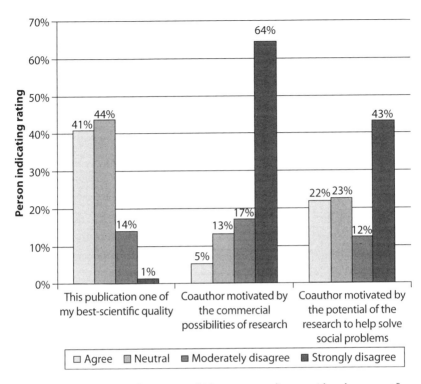

FIGURE 6.5. "Please indicate the extent to which you agree or disagree with each statement." (Number of observations = 629)

The distributions of levels of agreement about scientific, commercial, and societal motivations of coauthors are quite different from one another (fig. 6.5). Scientific motivations (represented by the label "This publication one of my best—scientific quality") received the fewest "disagree" ratings.[8] In contrast, motivation by the commercial possibilities of the research was strongly disagreed with by 64 percent of respondents. This tells us that while most coauthors of scholarly papers are not motivated by commercial pursuits, a few papers have academic researcher coauthors who are motivated by commercial possibilities. Motivation by social problems falls between these two poles. Twenty-two percent of respondents agreed that a coauthor was motivated to solve social problems. This comports with previous findings suggesting that most STEM researchers are chiefly motivated by the research puzzle, "the science," not contributing possible solutions to social ills (Ladd and Lipset 1972; Sarewitz and Pielke 2007).

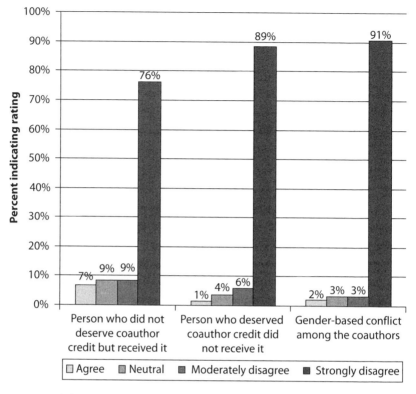

FIGURE 6.6. "Please indicate the extent to which you agree or disagree with each statement." (Number of observations = 627)

WHO IS DESERVING? COAUTHORS' PERCEPTIONS

The most important finding is that most respondents felt that in their most recent collaboration no one who should have been credited was denied and no one who should not have received credit was rewarded (fig. 6.6). But the results do indicate that dissatisfaction is certainly not uncommon, with nearly one-quarter indicating that someone who did not deserve credit received it and more than 10 percent thinking that someone who did deserve credit did not get it. Of course, as to the latter we did only survey the credited coauthors, not the ones who were left off, so we would probably get a different rating if we were to query those individuals who viewed themselves as collaborators but who were left off (notwithstanding the difficulty of identifying people left off a particular article).

If we take the "agree" and "neutral" ratings together, about one-fifth of respondents reported some type of problem with the most recent

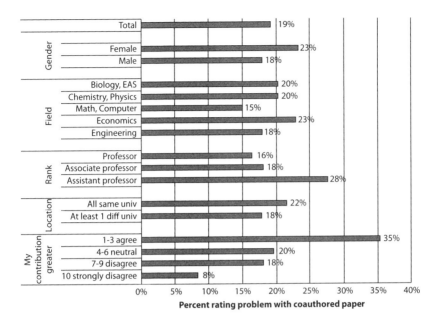

FIGURE 6.7. Coauthorship problems (measured by ratings of 1 to 6) by author characteristics.

coauthored publication. These ratings indicative of a potential problem with the coauthorship were higher for female than male coauthors, less common for those in math and computer science, higher for assistant professors than for associate or full professors, and slightly higher if coauthors were all at the same university than if they were at different universities.

There were also respondents who said their contribution was greater than their coauthors, and these researchers were more likely to report a problem with the collaboration (fig. 6.7). Thirty-five percent of respondents reporting that their contribution was greater (by assigning a rating of "one, two, or three") also indicated that there were problems with the collaboration, while only 8 percent of those who strongly disagreed with the statement about making a greater contribution also indicated a problem.

Assessing Research Collaboration Problems during the Whole Career

Now we turn to collaboration experiences over the academic researcher's entire career. Survey participants were asked to indicate how often four types of experiences occurred over the entire career: (1) a coauthor did not finish agreed-upon research-related activities; (2) coauthorship credit was

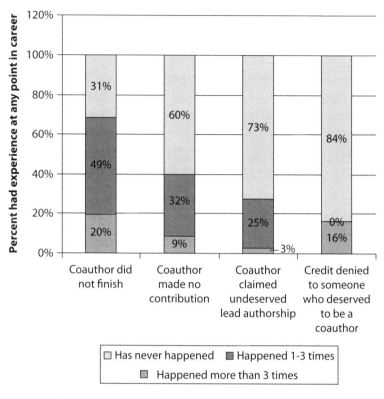

FIGURE 6.8. "Have you at any point had such [negative] experiences in your research collaborations?"

denied to someone who deserved to be a coauthor; (3) a coauthor claimed lead authorship when it was not deserved; (4) a person listed as a coauthor made no contribution at all to the research.

The most common negative experience is a coauthor who did not finish agreed-upon research activities (fig. 6.8). More than seven in ten respondents report this negative behavior at some point in their career. Next, but less prevalent, is that a listed coauthor made no contribution at all to the research—an issue mentioned by just over 40 percent of respondents—and 30 percent of respondents report that a coauthor claimed undeserved lead authorship. Least common is denial of credit to someone who deserved coauthorship, with only 16 percent of respondents indicating this negative behavior. Notably, among those who experienced these negative outcomes, they did not occur often but rather only one to three times over the career. This infrequency suggests that once respondents have a bad experience with a coauthor, they probably did not work with that coauthor again. Indeed,

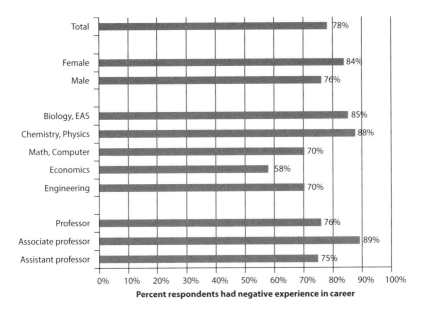

FIGURE 6.9. "Have you at any point had such [negative] experiences in your [research] collaborations?" (By respondent characteristics)

this strategy is borne out in some of our interviews with coauthors. The exception to this rare occurrence of problems concerns denial of credit, which, if it happened at all, is likely to have happened repeatedly.

Overall, 78 percent of respondents have had any of these four negative behaviors occur one or more times during their career (fig. 6.9). Female researchers are more likely than their male counterparts to have had these four negative behaviors occur during their career. Negative experiences are more common in biology/earth and atmospheric science and in chemistry/physics than in the other fields. These gender and field differences are consistent across all four types of negative collaboration.

Less easily explained is the greater likelihood among associate professors to have had negative research collaboration experiences than is the case for full or assistant professors. However, if we break this finding down across the four types of negative experiences, associate professors are more likely to have had a problem with undeserved lead authorships; 38 percent of associate professors experience this problem at some point in their career compared to only 24 percent of full professors and 28 percent of assistant professors.

In a separate analysis using this same dataset, Youtie and Bozeman (2014) find that collaboration problems can be mitigated by the decision-making process used. The results show that explicit authorship discussion significantly reduces the likelihood of problems. The same model indicates more problems with coauthors in the home institution than in a distant institution. In addition, academic researchers with a past bad experience are more likely to have had problems with their recent collaboration as are early career researchers.

Conclusion

We began this book by noting that most research collaboration experiences are positive ones and, indeed, the questionnaire-based findings, like our interview findings, bear that out. Coauthoring and collaborative problems with any given paper are unlikely, existing in any form in only 19 percent of respondents' recent papers. But few researchers manage to dodge the bad collaboration bullet for their whole careers; more than 80 percent of academic researchers reported at least one major negative research collaboration experience during their careers. When we consider that some of our respondents are young researchers, reporting on brief careers, it seems likely that bad things will ultimately happen, though in most cases small bad things. Researchers should expect to at some point deal with negative research collaboration experiences.

We noted that only 40 percent of respondents had explicit discussions about coauthor crediting relative to the most recent paper, even though this strategy is associated with fewer problems. We feel that the fact that the majority of collaborative teams never have explicit discussions of crediting explains much about unsatisfactory outcomes, and our interview data provide support for this view. In the final chapter we focus on a major element of the Aggregate Effectiveness Model, research collaboration management, the one element that so often provides team members the ability to take steps to improve their collaboration outcomes.

7

Enhancing the Effectiveness of Research Teams

THE "CONSULTATIVE COLLABORATION" STRATEGY

This book has covered the sometimes severe, sometimes mundane set of problems that scholars can expect in the course of collaborating on research. In chapter 2 we examined collaboration problems through a typology distinguishing routinely good, routinely bad, dream, and nightmare collaborations. We suggested in chapter 2 and documented later with various data sources that some research collaborations are remarkably successful (Dream Collaborations) but these are relatively few on a percentage basis and, fortunately, Nightmare Collaborations, ones that are extremely negative and damaging are quite rare. The vast majority of research collaborations are well and accurately labeled as "routine," either routinely good, meaning that outcomes were about as expected, or routinely bad, meaning that the collaborators experienced important glitches but not far beyond the bounds of normal expectations.

Despite that fact that most research collaborations have good or at least satisfactory outcomes, every collaboration entails uncertainty and even risk. Resources, reputations, good will, and scientific results are at stake in every collaboration. The question, then, is this: What can you do as a research collaborator or team leader to set your team on course for an effective collaboration? Nearly as important: When things are going wrong, what can you do to right the ship? During most of the book, we have spent only limited direct attention on the "what can you do to improve effectiveness" question. True, much of the book has presented cautionary tales and best practices anecdotes, but in this chapter we take on effectiveness issues directly and with specific prescriptions. We begin with an overview of the institutional context within which researchers collaborate, including particularly institutional review boards and various programs or codes advocated or required by government agencies or professional organizations. In most cases, collaborative teams have only limited ability to manage these institutional constraints, but since they provide frameworks and sometimes tools for collaboration, it is at least useful to review the institutional context.

We turn next to the Aggregate Model of Research Collaboration Effectiveness presented earlier in the book and distill lessons from that model, focusing particularly on the various components of effectiveness identified in research studies, our own and others. Finally, and most important, we identify and advocate a "Consultative Collaboration" strategy for managing research collaborations, a strategy we feel has significant potential for forestalling negative outcomes and increasing the likelihood of positive ones. Consultative Collaboration emphasizes team participation, open communication and alignment of values. Consultative Collaboration certainly cannot forestall or remedy all major collaboration problems. However, the approach does have the merit of focusing on activities that are clearly in the hands of the team members.

We are not inventing the Consultative Collaboration strategy from whole cloth, but rather we have observed the approach, and we have derived it from interviews and from our survey data. However, it is certainly not the most common approach to research collaboration management, and thus we develop a typology that accounts for most such approaches, comparing and contrasting their respective elements.

In the next section we consider the institutional context and particularly the institutional prescriptions that have been urged by others. In some instances we comment and supplement these recommendations.

Institutional Prescriptions for Improving Research Collaboration

When research collaboration problems develop, research team members do well to remember that there are institutional approaches that aim to help (and sometimes do!). There are existing pathways of ethics codes, training programs, review boards, and journal requirements, among the many approaches that merit discussion. These existing pathways often provide alternative means of resolving contributorship disagreements, ones that bear consideration before deciding upon the "nuclear options," such as involving the funding sponsor or the publisher in the coauthorship dispute, more drastic options that can affect all team members' ability to obtain future grants or research article placements. Yet almost none of our interviewees mentioned these existing codes, programs, or organizational approaches. Why is this the case? Possibly these norms and requirements are internalized to such a degree that little thought is given to them. More likely, their influence is not nearly as strong as the personal attitudes and group dynamics of research team members.

THE OMBUDS

The ombuds is an institutional resource found at most universities and deals with conflict resolution issues including but also expanding beyond research collaboration conflicts. The ombuds does not advocate for the researcher in a grievance, rather this person or office listens and, refers the researcher to resources at the institution, lays out options for resolution, and organizes mediation. Existing studies (Harrison 2004; Spratlen 1995) report limited effectiveness of an ombuds, in part because success is in the eye of the beholder (e.g., the disputant, the ombuds, or the institution) and in part because the ombuds process covers a very broad range of third-party resolution positions marked by conflicting goals, making success difficult to define and measure. Given the cross-institutional nature of many collaborative research projects, the ability of a single institution's ombuds to have an influence on these types of disputes is limited at best and typically not relevant to multi-institutional research collaborations.

INSTITUTIONAL REVIEW BOARDS

Much better known to researchers, the institutional review boards (IRBs) are another organizational source with the potential to improve research collaboration. The IRB primarily uses national standardized training and certification in research ethics, both online and in-class, and reviews of submissions explaining how the research will protect human subjects. For every research project, a university's (or other research institution's) IRB reviews and certifies that research methods following IRB standards for human subjects protections and ensures that conflict of interest issues are addressed. The national, if not global, reach of IRBs (which have different names outside the United States, particularly "science ethics boards") has the potential to influence research collaboration problems, yet none of the survey respondents or interview subjects we talked to mentioned the IRB, or indeed any university policy, as being an effective strategy for dealing with research collaboration problems. IRBs have been criticized for expansion of rules beyond original objectives, lacking relevant research training for decision making, adding bureaucratic requirements and delay, and applying an ever-growing number of rule-based techniques that do not necessarily protect human subjects (White 2007; Millum and Menikoff 2010; Bozeman et al. 2009).

"TOOLKIT" APPROACHES

In response to perceived weaknesses in existing code-based, program-based, and organizational solutions, we have observed a rise in "toolkits" to improve effectiveness or to provide solutions to research collaboration problems. For example, two researchers at Florida Atlantic University developed assessments to address personality-based issues in collaboration; the Collegiality Assessment Matrix and Self-Assessment Matrix ask about faculty member behavior relative to an external group or to the particular professor (Schmidt 2013).

The Science of Team Science apparatus of the NIH's National Cancer Institute provides an especially elaborate toolkit approach. Their toolkit offers research papers, training materials, introductory agendas for new research projects, needs assessments, model agreements, surveys, and metrics and scoring methods. The toolkit has an orientation toward biomedical research projects but could have applicability to other types of research that involve large numbers of research project team members.[1]

While toolkit approaches have their place, there are several issues with overreliance on them. First, use of assessments, questionnaires, and training as a matter of course can add another layer of bureaucracy and paperwork on top of researchers' existing administrative burden to provide monthly reports, export control forms, conflict of interest forms, IRB forms and certifications, certifications associated with use of proprietary data, annual reports, and the like. Second, these assessments, questionnaires, and personality tests have the potential to be used to tamp down on dissenting views, which are an important part of the research process (as we shall see below in the discussion of Consultative Collaboration). Third, research problems can be idiosyncratic, particularly those of smaller groups, and more difficult to reduce to a single solution.

TEAM-FOCUSED PRESCRIPTIONS VS. TOOLKITS

If there is anything that this book shows, it is that there is no one best way to resolve research collaboration problems. If you are looking for a toolkit, this book is not it. Management science approaches in the early 1900s were found to not work well for manufacturing tasks (Schachter 1989; Wrege and Perroni 1998), and they are likely to be even less useful for managing research. Indeed, even manufacturing itself has moved away from standardization and hierarchy toward more participatory approaches through cellular manufacturing, quality circles, and self-directed and cross-functional teams. Although some have made analogies between the rise of large-scale science and assembly line manufacturing (Walsh and Lee 2015), we would expect that standard toolkits would be even less useful in managing research than flexible and participatory approaches.

One reason is that some of the barriers to effective research collaborations are intractable. There will always be gender differences, power asymmetries between researchers at different career stages, and differences in the way individuals fit into a research team. A second reason is that the causes of nightmare collaborations are many and complex. As we indicated in chapter 4, what might evoke an extreme negative reaction in one case might be viewed as "business as usual" or even as extremely beneficial in another case. Returning to our example from chapter 1, having a famous scholar attached to a paper who has done no work on it—i.e., an honorary author—may cause major problems for one author who may have made a contribution to warrant authorship but was subsequently displaced as the first author by this famous scholar, especially if the displaced author was at an early stage in his or her career and really needed the first authorship. Another author whose paper was able to gain access to more prestigious journal placement with a wider audience as a result of the famous scholar's being the first author may view this same situation positively. And yet another author may not give the inclusion of an honorary first author another thought because this kind of crediting is the norm in his or her culture.

So what can be done to address research collaboration problems? It is not useful to oversimplify collaboration problems, as we have seen in the aforementioned toolkits, but it is also not useful to treat every problem as idiosyncratic. To this end, we take a two-pronged approach. First, we return to the aggregate effectiveness model presented in chapter 4, reviewing effectiveness factors. Second, we provide at the end of this chapter a set of archetypal approaches to management collaboration, arguing for Consultative Collaboration as a means to building effective research teams.

Prescriptions from the Aggregate Model of Research Collaboration Effectiveness

Our aggregate model considers research team effectiveness in terms of four internal factors—Career Stage, Trust and Experience, Work-Style Fit, and Gender Issues—and external factors concerning discipline and sector. Each factor is presented here using one or more illustrative problems (table 7.1). To add context to these problems, we include representative quotes from submissions we collected from survey respondents who offered examples of negative (as well as positive) research collaboration stories. As we discussed in chapter 1, these stories were submitted as part of the Web survey, but in the form of posts on a separate and anonymous Web page that was not part of the formal Web survey form. After presenting these problems and posts, we discuss strategies (summarized in table 7.1) for addressing each problem and risks associated with pursuing each strategy.

CAREER STAGE

Career-related coauthorship problems are traditionally viewed as being caused by power asymmetries between senior and junior researchers. But the literature does not always support this career-stage dynamic, as we saw in chapter 5. Indeed, some studies find higher productivity among junior researchers or no career stage difference. Most of our interview and survey data highlighted career stage differences in the expected direction, however:

> I had collaborated with a colleague on a previous study which had been published about a year ago. We did a follow up study, using the knowledge that was published in that previous study as a starting point to design new experiments, but with no further involvement of this colleague. When I prepared the manuscript for submission, that colleague insisted that she and her student be added to the paper, while they had not contributed to any of the work. When I replied that I did not think it

Continued on next page

TABLE 7.1. From the Aggregate Model: Problems, Strategies, and Risk

Factor	Problems	Strategies	Risk
Career stage	As a student or junior faculty member, you do all the work on a paper and your advisor or senior colleague adds him/ herself or puts him/herself in the most advantageous position in author order.	Present a reasoned argument about the preferred author list/ order to the advisor/senior faculty. If not successful, escalate to the chair of the department. Make sure that anytime the advisor or senior faculty asks about author order that you (as a student/junior faculty) always indicate that you should be in the author position that you deserve and even follow personal conversations with an email to confirm the relationship. This is no time to be modest.	Resolution puts the burden on the student/junior faculty member, who is at a disadvantage in the power relationship.
Career stage	Two students make the same contribution to the research, so you cannot choose which should be first author.	Investigate whether the journal offers an option that enables two first authors. If not, try to carve out two studies that enable each to be the first author on at least one paper.	One student is likely to be disadvantaged.
Career stage	Your student is not capable of doing the work on the project.	It is up to you as the senior member of the team to proactively contact the student. Find out what the student can do, then either find another researcher (e.g., another student, postdoc, colleague) to do the other aspects of the research	The student misses a learning opportunity.
Trust and experience	Coauthor insists on adding someone to an article you wrote and on which you are the lead author; this added person made no contribution to the work	Discuss with the coauthor. Make a reasoned argument why the person should not be included as a coauthor. If this does not work (because the added person is too senior or some other reason), carve out another research question from the data that does not involve the added person.	The involvement of people who cannot stand behind the research. Authorship of those who did perform the research may appear to have less value depending on the order of the uninvolved coauthor on the publication.

TABLE 7.1. (*continued*)

Factor	Problems	Strategies	Risk
Trust and experience	Foreigner includes you as a coauthor without asking you and the paper is published.	Let the foreigner know that you do not wish to be a coauthor and ask to be removed in an addendum or errata to the paper if you really do not wish to be part of the paper. Otherwise, just let it go and make sure that there is a formal understanding about your author inclusion policies in any future foreign researcher relationship.	If the foreigner's research is not up to your standards, that research is forever associated with your name.
Trust and experience	You give data to another researcher who does not include you as a coauthor	Do not work with that author again. Make it a policy to include coauthorship stipulations in future data sharing arrangements.	Missing out on that author's potential contribution in the future.
Trust and experience	You need a minor amount of data or samples from a researcher who insists on substantial coauthorship credit	Look for another data or sample source or try to negotiate another form of credit with the researcher.	You are not able to access important data or samples.
Work-Style Fit	A senior colleague creates a lot of interpersonal problems on a research project.	Get through the project as best you can, and then do not work with the senior colleague again on future projects.	Missing out on that author's potential contribution in the future. Lack of communication fails to resolve issues.
Work-Style Fit	Coauthor will not have an explicit discussion about authorship crediting at the beginning of the project.	Indicate which papers you want to lead in a telephone call or meeting and follow up in an email.	Research may follow a pathway that is difficult to foresee in advance.
Work-Style Fit	You cannot get the coauthor to do enough work on the project.	Assign that work to other coauthors and yourself. Demote or remove coauthor from author list and communicate this directly to that person.	The performing coauthor might have specialized knowledge that other members cannot take up.

Factor	Problems	Strategies	Risk
Gender	Female postdoc goes on maternity leave and is left off publications stemming from projects she spearheaded.	Have a discussion with the supervisor in advance of maternity leave about authorship inclusion while away. If exclusion persists, find another group to work with on returning from leave.	The solution punishes the female postdoc for having a baby.
Gender	Female is negotiated out of prominent author position.	Have an upfront policy about what constitutes authorship position when working with colleagues.	The level of formality may be perceived as too awkward and alienating and puts off some collaborators.
Discipline	There are too many people involved in a project to be coauthors on every paper.	Lay out a list of papers at the beginning of the project, ask who wants to work on a set of papers, balance interest with skills, and let this process guide assignments	Papers that result cannot be predicted at the beginning of the project
Discipline	A technician did critical work on the project, yet is not included as an author.	Establish requirements for authorship and appearance in acknowledgments. at the beginning of a project.	A novel and influential contribution by a technician is not able to be recognized because of authorship policy.
Sectoral	Coauthor suppresses results because of conflict with intellectual property, company profitability, or standard operating procedures.	Use conflict of interest, editorial policies.	These methods will be ineffective without better training and socialization.

was fair, that colleague (who was a senior member of my dept.) replied to me that "this wasn't the smartest thing to do for your tenure promotion". I ended up adding both people to the paper.

This quote suggests that the junior faculty member has few options except to accede to the senior faculty member's coauthorship demands. If this were the case, it would not be surprising that junior researchers were to leave the field, which is what we conjectured in chapter 5. However, as one faculty member indicates below, it is extremely important careerwise to make sure that early publications fairly indicate authorship:

Before tenure worked with a more senior person. . . . The senior person had expertise in a completely different field, so the measurements and analysis were completely guided by myself. The paper was written, but my name was buried in the middle of the list, rather than near the end, which signifies supervisor in my field. When this was brought up with the senior colleague they blew up. Initially backed down, but as promotion loomed, made an additional request. Colleague blew up again, but did make the change. Moral: junior faculty, if you foresee that the majority of the work is occurring in your lab, you need to be proactive and upfront about author order BEFORE paper is written. I have never again collaborated with this faculty member.

In cases such as the above, we recommend a proactive strategy in which the student or junior faculty member makes a logical case for giving the preferred author position to the more senior author. This preferred position is usually the first author for a student or the last author for disciplines in which the supervisor or laboratory director typically appears at the end of the author list. Sometimes the senior faculty member will ask the student/junior faculty member whether he or she thinks the senior researcher's contribution warrants a particular authorship position; the student/junior faculty member should be ready to be honest in advocating for the preferred position for themselves and a lesser position (or even no position) for the senior faculty member. Moreover, the interaction should be followed with an email to confirm any agreed-upon contributorship outcome.

It is best if the student or junior faculty member is able to resolve the authorship dispute among the affected parties. However, the student/junior faculty member should be prepared for the fact that the senior author may not yield the preferred authorship position. If all disputing parties are in the same university and department or school, the junior researcher should take the case to the chair of the department. This case will need to be more formally documented, such as through a brief timeline, justification, and any relevant emails.

Let us provide one comment about graduate students. Faculty members will sometimes have to navigate situations concerning which of their students made the greatest contribution on a single paper and thus deserves first authorship. They also will have to deal with students who are not fully aware of the ethics of authorship on papers they submit. For example:

> On one occasion, a graduate student believed she should be first author, whereas I disagreed. Ultimately, I designated that the first and second authors (both graduate students) had made equal contributions to the paper.

and

> I have a PhD student who I inherited from a professor who was denied tenure. He is strong technically but has some really poor training related to research and publishing ethics. He recently submitted a paper we worked on together without telling me about it until after the fact. The paper had at least 70% overlap with another paper we recently finished. . . . We pulled the paper and thankfully he will graduate this quarter. Not a bad guy, but really weak on some important research issues. I think if I'd worked with him from the beginning, I could have trained him properly but now it seems to be too late.

In these instances, it is up to the faculty member to mediate authorship disagreements between research assistants or postdocs. Asking the students or postdoc to resolve authorship disagreements themselves may work, but it also may exacerbate problems. Where the disagreement involves a first-author dispute, the senior faculty member or laboratory director should contact the journal to find out whether the journal offers an option that enables two first authors, such as through the use of an asterisk and explanatory note (Venkatraman 2010). The growth of co–first authors, especially in the biosciences, has led to calls for greater coordination across journals in developing a standardized method for designating co–first authors such as through the use of bold lettering (Conte et al. 2013). As yet, though, co–first authorships are not widely recognized, especially outside of the biosciences. In lieu of this option, the senior investigator should work with the team to carve out multiple studies in which junior researchers would alternate taking the lead author position.

Likewise, it is the responsibility of the senior faculty member to provide ongoing research ethics training. Socialization and training is key to the creation of fair and ethical research practices. Of course, most federal agencies in the United States require IRB certification to work on sponsored research grants. Responsible conduct of research (RCR) training is required for participation in NSF research grants as called for in the America Creating Opportunities to Meaningfully Promote Excellence in Technology, Education, and Science (COMPETES) Act,[2] and NIH has requirements for RCR training as well. But training does not always translate into practice. It is up to the senior member of the team to proactively reinforce the ethical practice of research, regardless of the student's career stage.

TRUST AND EXPERIENCE

As we indicate in chapters 4 and 5, collaborations with productive research-ers who make a distinctive contribution over a long period of time engender the most trust. Even though the concept of trust appears straightforward, research contributorship incidents are much less so. For example from our survey:

> I had someone that was competent but desperate for publications claw their way onto an authorship on a multi-university, interdisciplinary study that I ran when this investigator had no part in the study but had a post-doc that had participated (this post-doc participated on her own time, on weekends since the lab director had refused to let her officially participate in the project). Once it became apparent that the paper would be published in a prominent journal, the post-doc's PI used inter-university politics to mandate their inclusion as an author. In 200+ publications, that is the only one that I regret in terms of authorship. However, I learned from it - I now NEVER collaborate with anyone I don't like and respect as a person.

A common strategy team members use to deal with trust and experience issues is to be extremely inclusive in listing authors, even authors who do not make much or sometimes any contribution to the work. This inclusive approach would seem to be the most reasonable way to decide authorship, because it avoids contentious interactions, but it is not always a good strat-egy. The reason for our recommendation against the reflexive use of the inclusive strategy is that some of those listed as authors may not be able to stand behind the work if it were challenged, especially if challenged through replication. A better approach would be to list authors based only on their contribution to the work and recognize others in the acknowledgements to the paper. Indeed, the majority of respondents to our survey said that they based authorship decisions on an assessment of each author's research contribution (refer to chapter 3). Contributorship requirements in some journals, especially in health and biomedical fields, now require coauthors to list their specific contribution. While there is not much evidence about the effectiveness of this approach, we expect that it has some utility. Persons who have done little or nothing on a research project might find it easier to passively accept a coauthorship than to lie in print about their role or lack of a role.

Another set of problems occurs when an author is included in a paper without the author's consent. This kind of unwanted authorship occurs especially with foreign researchers who may have different norms for including and informing coauthors. One of our survey respondents reported a case of unwanted authorship thus:

> I discovered that I had been listed as the first author on a publication of which I had no prior knowledge. The publication was of poor quality and I was quite unhappy to be associated with it.

In this case, the unknowing author let the publication stand. We recommend, however, that the unknowing author pursue the option of having his or her name be removed in an addendum or errata to the paper. This option should be sought particularly if the unknowing author cannot stand behind the paper's findings. If there is no recourse but to publish the paper, the unknowing author will have to accede. Nevertheless, the author should make sure that an explicit agreement about his or her inclusion policies be communicated in future collaborative relationships.

WORK-STYLE FIT

Work-style fit is not just a matter of harmonious personalities. As we saw in chapters 4 and 5, it really concerns collaborators working at similar paces (steady vs. erratic) and with similar styles (rigid vs. flexible). Illustrative of work-style fit problems are the following nominated experiences from our survey:

> I have a collaborator who NEVER reviews our work or prepares for meetings. We always have to start over at the beginning because he has no idea

what is going on. He then doesn't read the papers and complains when we send his name further and further down the list. But, he has more research funding than me, so he has more power in the department and he will get promoted to full faster than me too.

Another respondent focuses on an inattentive collaborator:

I did have a co-author (when I was untenured) that repeatedly ignored my requests for input on a manuscript. He provided little help except for providing some supporting data. Feeling pressure to publish, I finally submitted the manuscript with his name included as a co-author. Several months later, when I informed him and the other co-authors of the papers acceptance (In a relatively prestigious journal), he wrote me a very negative letter and berated me for being so unprofessional and submitting a manuscript without his approval (he was also untenured, but a few years senior to me). Being told I had acted in an unprofessional manner so early in my career upset me greatly. I wrote him back and told him that I would withdraw the paper, take his name off, remove his data and re-submit the paper elsewhere—to a lesser journal since his data were helpful. He responded back that he would prefer I did not follow this course of action—and that this was simply a misunderstanding that we would laugh about someday. I did not collaborate with him again.

These problems illustrate the difficulties of working with a colleague who could make a useful contribution but, for whatever reason, does not work at the same pace as the rest of the team. Moreover, in both cases, the offending researcher has style-related issues with the pecking order of the project. In the first case, the collaborator resents being further down the list. In the second case, the collaborator does not respond to requests for input until the paper is accepted, and then sends a harsh letter admonishing the lead author for not waiting for his input.

In general, we recommend that the lead author deal with work/fit issues through explicit communication early in the project and regularly throughout. If punitive measures are required to contend with nonresponsive colleagues, the lead author should be prepared to reassign this person's work or reconfigure the project without this person. The caveat to reassignment is that a candid assessment of lateness should be conducted before taking such action. The lead author should ask whether the deadlines are reasonable or too optimistic and whether a delay will dramatically hurt the project. It is easy for a lead author to take a delay to heart (especially when the delay is holding up a publication much needed for promotion or tenure), when in fact the delay is due to the offending researcher's personal issues or erratic way of working and is not meant as an attack on the lead author. Sometimes a well-chosen meeting to iron out scheduling differences and revamp deadlines will work wonders. Of course, some authors have such trying personalities that the explicit approach will only cause problems for the lead author. In these instances, the lead author should try to get through the project as optimally as possible and then avoid working with the offending colleague in the future.

GENDER

Previously we noted the difficulties in disentangling issues caused by physical factors relating to gender (such as maternity leave) from those related to social and cultural contributory concerns. Concerning the latter, as one survey respondent described it:

> The ways males and females negotiate authorship seems different. I have had male co-authors assertively state what they apparently feel they deserve, while female co-authors (including myself!) appear to value smooth relationships (or simply to avoid conflict) so much they are willing to take a back seat in terms of authorship, even when "technically" that should not occur.

In addition to these factors, gender often exhibits strong interaction effects with other internal and external factors, as we emphasized in chapter 5, particularly interactions with career stage. Women tend to be in earlier career stages and junior positions in research collaborations. Thus, while coauthorship positions may be in a precarious position due to maternity leave, positions may also be precarious due to socially determined approaches to negotiating these positions.

We suggest that women in STEM disciplines should make particular use of upfront discussions in advance of the research. These discussions, before maternity leave or at the beginning of a research collaboration, have the potential to help female scientists retain their coauthorship position. While the risk is that some collaborators may be put off by such a stilted manner of working together, as well as the previously mentioned lack of being able to predict how the science will turn out, fairness in coauthorship decisions is still enabled by upfront discussions. Moreover, these decisions can be subsequently modified if major changes in research outcomes occur.

We strongly advise women researchers to avoid assuming that gender-based collaboration can be avoided by simply working with other women. While some women researchers have strong empathy and consideration for women colleagues' concerns, our results show that in some cases women, just like men, have a tendency to think, I had to go through these problems, what's so special about her?

DISCIPLINARY DYNAMICS

We made broad reference to differences between fields that require collaboration and those with more of an individual orientation to research. Some types of problems are also more likely to occur in certain disciplines than others. For example, disciplines such as high energy physics or biomedical research can have fifty or more authors (Cronin 2001). Sometimes all these authors are not always able to be included, especially in a printed journal article. One reason is that not all of them actually contributed to major research tasks: conceptualizing the idea, performing the research, or writing results. Here is a Web post that exemplifies this issue:

> I find working with individuals on the medical side a bit frustrating. An NIH proposal can include almost 15 co-authors (some are co-PIs, some are not). When writing a paper that is based only on one task (contributed by perhaps 4-5 individuals at the most), they all insist on being a co-author even if everyone did not contribute.

Another issue has to do with access to discipline-specific instruments or samples. To what extent does owning a piece of equipment or providing a standard reagent qualify for authorship? Consider the following:

> A person who had control of an instrument that we required to run experiments refused to allow access (even though it was owned by our institution, not this person) unless he was granted lead author on our next publication, in addition to co-authorship on the one we were currently working on. This person did not participate in the design of experiments, writing, or discussions, and did not provide any funding for the instruments or supplies, but was the supervisor of the person that trained others in how to use the equipment.

There is certainly no magic bullet to dealing with disciplinary issues. When people come together who have had very different training, perhaps different professional norms and, importantly, who have their professional stakes in the ground of a particular discipline, perfect harmony should not be expected. One modest suggestion: disciplinary issues should be dealt with at the time the collaboration commences. In particular, rules for who should or should not be an author should be established. As we know, there are considerable differences with respect to contributorship, and what is standard behavior in one discipline may be consider borderline unethical in another. The "up front deliberation" approach seems intuitive, but it is not always able to be followed, especially when team members are new to interdisciplinary collaboration and may not have yet developed an appreciation for the many possible pitfalls.

SECTOR

Collaborations with the private sector have always been beset with complications. But one of the most worrying with respect to contributorship is suppression of results. Suppression of results can occur when findings are not published because they conflict with the expected results, intended application, planned research agenda, or desires to protect an invention for financial benefit. It is essentially a struggle between research and corporate profitability interests. We think of suppression of results occurring when one member of the research team works in academia and the other works in industry. However, trends supporting encouragement of entrepreneurship at universities (Shane 2004) can lead these potentially diverging interests to be held by the same person who is working on a research project and also starting a company based on that research. One respondent to our survey described just such a collaboration:

> One negative collaboration occurred because one of the co-PIs [from a different university] did not want to publish the results, which may have been influenced by IP reasons and his startup company.

The ability of academic researchers to publish their work is typically protected in university research agreements. These agreements also include provisions allowing for company review of the publication in advance and assurances that company proprietary information will not be revealed in the publication. University conflict of interest policies also prevent institutional researchers from having their work influenced by financial benefit in a startup company because such conflicts have to be reported as a condition of employment. Likewise, some journals have editorial policies that require financial conflicts to be reported. Whether these policies are always complied with is another story, though legal ramifications from the university and retractions from the journal can result if they are not. Better training and socialization would make the existing protections even more effective in preventing suppression of results stemming from difficulties in negotiating scholarly and entrepreneurial interests.

Toward Consultative Collaboration

If we consider all the collaboration problems researchers face and all the approaches used to either avoid problems or to resolve them, then our data tell us again and again that the team's collaboration management approach is a vital ingredient in research team effectiveness. Indeed, the *science* of team science is largely *management* science. In reflecting on all the data we have developed for this study as well as findings from other studies in the S&T Human Capital project, we can distill characteristic approaches to managing research teams. Here we present a typology of management styles we have observed, providing an assessment of each. We strongly prefer the research team management approach we term Consultative Collaboration as the approach most likely to lead to research collaboration effectiveness and, if fully embraced, the approach that is nearly guaranteed to help avoid Nightmare Collaborations.

The research team management approaches we describe should be viewed as "archetypes," meaning that they are built on the most common, modal characteristics. But in actual practice there is some overlap, and any particular approach may not manifest each and every characteristic we associate with the archetype. With that in mind we identify the following collaboration team management approaches we have observed, listed below in ascending order of preference. The research team management approaches include:

1. Tyrannical Collaboration Management
2. Directive Collaboration Management
3. Pseudo Consultative Collaboration Management
4. Assumptive Collaboration Management
5. Consultative Collaboration Management

Table 7.2 elaborates on these management approaches, providing key characteristics, a rough estimate of their proportions among all research collaborations, and a brief comment on the effectiveness of each approach.

TABLE 7.2. Research Collaboration Management Archetypes

Collaboration management type	Core components	Proportion of all collaborations (authors' estimate)[1]	Effectiveness?
Tyrannical	One person is aggressively in charge and has a lack of interest in a lack of respect for other collaborators' values and preferences. Makes all key decisions about collaborator choice, task assignment, knowledge disposition, crediting.	10–20%	Almost always undermines the collaboration, causes resentment, reduces likelihood of continued collaboration.
Directive	Like Tyrannical, one person is in charge of the collaboration and makes key decisions. However, not malevolent or	20–30%	Not necessarily effective or ineffective, depends on skills of the Directive and other collaborators' assessments of those skills.
Pseudo Consultative	A simulacrum of consultation but cues suggesting those some collaborators' values and preferences are less important than others.	5–10%	Not necessarily effective or ineffective; can be harmful if the persons in charge are manipulative.
Assumptive	One or more collaborators simply assume others' agreement about preferences and values for the collaboration. The assumptions may or may not be correct.	50–60%	Can be effective when assumptions are based on trust and previous experience, otherwise generally ineffective.
Consultative	All parties to the collaboration are consulted at key points in the life of the collaboration in order to identify respective preferences and values and decide upon specific actions.	5–10%	Generally, the most effective approach to collaboration management. Key is degree and timing of consultation (otherwise can levy heavy transactions cost).

We discuss each of the research team management approaches below, but it is useful to provide additional caveats about the typology. As mentioned, these are archetypes representing characteristic behavior but not collaboration behavior in all its great nuance and diversity. Nevertheless, we feel the typology captures the basic aspects of team management and accounts for a very large percentage of collaboration approaches, well in excess of 90 percent of the collaboration management approaches we have observed. A second caveat: some research collaborations are so ill managed or unmanaged that they cannot be said to have a management approach at all. Sometimes management simply is neglected and chaos reigns supreme. We do not include "no management" in our typology, but we come back to this point below. Third, while most collaborations employ only one basic management style, some of them evolve, moving from one style to another. Usually this change of style comes in response to problems that arise. However, it is also common to have Assumptive Collaboration Management as an "end state," as it is in some cases an approach that develops from experiences with particular individuals that, over time, engender not only familiarity but also trust.

Below we review each research collaboration management approach in turn.

TYRANNICAL COLLABORATION MANAGEMENT

Almost invariably the worst of management styles (except perhaps for "no management"), Tyrannical Collaboration Management features a strongman (-woman) approach where "collaboration" is essentially a euphemism. Within the research team, or in other team settings (e.g., Ashford 1994; Kant et al. 2013), Tyrannical Collaboration Management occurs when one person is aggressively in charge and has a lack of interest in or a lack of respect for others' values and preferences. In most cases, not all, the Tyrannical Collaboration Management is attributable to the personality flaws or ruthlessness of a single individual, often someone who is either especially eminent in his or her field, is more productive individually than the other collaborators, or who controls resources that other collaborators depend upon. Often tyranny occurs as more senior researchers seek to dominate junior researchers, postdocs, or graduate students. In most cases the perpetrator concludes, usually with no evidence and certainly without the assent of others, that he or she is more able than the other collaborators. This extreme self-assurance may even extend to berating other team members.

The tyrannical team manager equates experience and power with wisdom, including managerial and decision-making ability. Since this behavior often happens in cases where there is a power asymmetry, these assumptions of superiority and entitlement may go unquestioned and the tyrant may take the lack of dissent as assent. In all likelihood, the tyrannical party will eschew collaboration with peers because persons with similar status and experience are not likely to long be subject to tyrannical collaboration.

In our observation, Tyrannical Collaboration Management is relatively rare, probably less than 10 percent of all research collaborations, but are especially likely to yield Nightmare outcomes, albeit ones that may never come to light, due to the fears of junior collaborators. Even when Tyrannical Collaboration Management does not lead to outright disasters, it usually results in long-term negative effects on motivation and team members' receptivity to collaboration. Often the worst effect of Tyrannical Collaboration Management is modeling for graduate students, the most inexperienced of whom may conclude that Tyrannical Collaboration Management is the norm and may even replicate bad practice.

DIRECTIVE COLLABORATION MANAGEMENT

In Directive Collaboration Management, an especially common mode of collaboration management, one person is in charge of the collaboration and makes unilateral decisions on most or all crucial issues, such as choice of team members, task assignment, knowledge disposition, and crediting. In most respects Directive Collaboration Management can be viewed as Tyrannical Management Collaboration Lite. Many of the behaviors are similar, including centralization of decision-making power, but the personality flaws and malevolence are missing or at least ameliorated. In some cases Directive Collaboration is a viable approach, whereas Tyrannical Collaboration never is effective, at least not in the long run.

Directive Collaboration Management in certain circumstances (Peterson 1997) can lead to positive outcomes for a team. For example, in teams composed of doctoral students and one highly experienced senior researcher, directive collaboration may be effective, especially if the team leader's behaviors model effective, experience-based, and benevolent choices. But in most cases, Directive Collaboration Management, though quite a common approach, is not a preferred approach. In most cases, even in the best-case scenario, where the directive research team manager is a person of good will and effective in most tasks, the other collaborators learn little about collaboration because they have little or no role in decision-making and usually no information about how the directive manager made decisions. The only learning that can occur is based on behavior modeling with inferences based on outcomes rather than direct observation.

Most instances of Directive Collaboration Management do not even approach the best case, the benevolent, effective single decision maker. If the directive research team manager makes bad decisions, then lower-status collaborators may feel reluctant to report their views, even if the team leader seems receptive. The younger, inexperienced members may be prone to question the conclusions they come to. As we know from years of study of group behaviors, one of the primary advantages of participative management and the "informal organization" is that team members may in some circumstances be able to save team leaders from potentially disastrous decisions (Spector 1986; Finkelstein et al. 2013). But such safety checks are unavailable if the team members do not even know about the team manager's decisions until after they have been made and implemented.

One of the reasons why Directive Collaboration Management is quite common is that the PI mode of work breeds such an approach because the authority in grants and contracts is often centralized in the PI. In our extensive case studies (Rogers 2001; Bozeman and Rogers 2001; Corley et al. 2006; Boardman and Bozeman 2006) of the management of government and university research labs and research centers, we determined that most directors gravitate naturally to their experience as PIs and that most such experiences are based on a single-leader model where one person is in charge of most decisions, and if there is consultation at all, it tends to be about issues specific to the scientific and technical content of work. Often there is a sort of path determinant approach where new lab directors rely on their experience as PIs, almost as a direct extrapolation, and then if they are appointed research center directors they usually bring the same directive collaboration approach learned as PIs.

Directive Collaboration is most likely to occur when all or most of the research collaborators work in the same lab or research center or, more rarely, when multi-organization collaborations include collaborative partners that are not fully invested in the collaboration and, thus, are often more than willing to have someone else take the managerial lead. Directive Collaboration Management occurs less often when PIs and team leaders have multiple roles and are engaged in organizational boundary spanning (Mangematin et al. 2014).

PSEUDO CONSULTATIVE COLLABORATION MANAGEMENT

While not common, Pseudo Consultative Collaboration Management approaches tend to be ineffective. In this management type there is the *appearance* of consultation, but in an environment providing cues that the values and preferences of some collaborators (e.g., cronies or the more powerful) are more important than those of others (e.g., outsiders, new team members, the least powerful). What is especially pernicious in the case of Pseudo Consultative Collaboration Management is that those who are in charge often are self-consciously manipulative, wishing to dominate the collaboration while not seeming to do so. If they are quite adept at such manipulation, then other parties to the collaboration may not have insight into the social dynamics involved (Galinsky et al. 2006).

The veneer of participation in what is actually an authoritarian, individualistic management approach is not unique to research collaboration and, indeed, seems more common in settings where political skills and the ability to manipulate are more highly developed (Riker 1986; Gardner and Martinko 1988). However, it is quite possible that management by manipulation is even more damaging in research teams than in other settings where political scheming and imagination are more common, tolerated, and sometimes even prized. Research collaborators do not generally expect to be manipulated, and thus a Machiavellian research manager may have an easier path to dominance and a greater ability to do damage.

ASSUMPTIVE COLLABORATION MANAGEMENT

Perhaps the most common approach to managing research teams, Assumptive Collaboration Management, occurs when some collaborators, often the most experienced or those in authority, simply assume others' preferences and values for the collaboration without actually establishing agreement and without providing an environment that makes it easy for other team members to provide values and preferences.

Assumptive Collaboration Management differs from Directive Collaboration Management in a couple of important ways. In the first place, under Assumptive Collaboration Management the assumption of others' preferences may be valid, especially in those cases where the collaborating parties have a long history of working together (Gruenfeld et al. 1996). After three or four collaborations, research colleagues often have a very good and valid knowledge of one another's preferences, values, strengths, and liabilities.

The key here is the validity of the knowledge (Dirks 1999); if the understanding of others' preferences and values is inaccurate, then wrong and harmful assumptions can perpetuate dissatisfaction. However, if all parties' assumptions are correct, then this particular form of collaboration management can be very effective because the transactions costs associated with research collaboration, including collaboration management but also the research tasks themselves, may be drastically reduced. When the team has "chemistry" and predictability, then research management may be optimally efficient. However, it is important to remember that the fact that collaborators have worked with one another for a long time or even the fact that they bring a high level of trust to the collaboration, does not invariably mean that the mutual assumptions about values and preferences are correct ones (Holton 2001).

A key element to effective research collaboration management is to bear in mind that circumstances change, and even when circumstances do not materially change, people change. While collaborators may view it as a waste of time to go over issues (e.g., author order, crediting, journal submission decisions, managing subordinates) that they have addressed many years ago, it is nonetheless the case that some revisiting of collaboration issues may be beneficial.

As might be expected, Assumptive Collaboration Management is the management approach practiced by long-time collaborators. If the approach is perpetuated then in most cases it is working. However, in some cases the continuation of Assumptive Collaboration Management may be owing to the fact that team members, many of whom may be introverts, wish to avoid even the potential for in-the-open disagreement or conflict (Hillier 1969; Colcleugh 2013).

Assumptive Collaboration Management often presents a special problem for research teams: a new collaborator just beginning in a project involving a collaborator combination of long standing may feel that she or he is a stranger in a strange land, bound by strong but unspoken or poorly articulated norms (Erikson and Gratton 2007). Assumptive Collaboration Management cannot easily accommodate the needs of new team members, at least not in the short term, and it is a good idea to directly and systematically accommodate new entrants (Klotz et al. 2014).

CONSULTATIVE COLLABORATION MANAGEMENT

We conclude with the gold standard for research collaboration management, Consultative Collaboration Management. In this style of collaboration management, all team members are consulted at key points in the life of the collaboration (formation, goal setting, task assignment, crediting, disposition and dissemination of intellectual property) so as to identify respective preferences and values and decide upon specific actions. While the factors below should not be thought of as absolute preconditions for Consultative Collaboration, they characterize Consultative Collaboration in its archetypal form:

1. Sufficient communication structures, used at key times
2. "Whole team" communication rather than communication among subgroups only
3. A high level of trust among collaborators (even if a new collaboration)
4. A weak or nonexistent association between the assessment of ideas and the individual presenting the ideas
5. A commitment to open and transparent disagreement, should there be disagreement
6. Revisiting of points of agreement and plans of action
7. Recognition of diverse values and objectives
8. Sensitivity to differences in status and power

Each of these points is worth discussing in turn.

Sufficient Communication Structures

Here the issue is not how often colleagues collaborate but rather the utility and quality of the collaboration. Indeed, too much communication can actually undermine the collaboration (Bartoo and Sias 2004; Madlock and Martin 2012). Communication requires time and energy, and if a researcher finds that an inordinate amount of time is being spent on communicating about small or trivial matters, then commitment to communication is likely to wane. Consultative collaboration involves *effective* communication, not *constant* communication. The media of collaboration also bear some attention. It is not a good idea to simply assume that "everyone prefers email" or that "we all use Skype," or even that face-to-face communication is best. It is important to understand the contingencies of communication, including not only personal preferences but also the most appropriate communication medium for the topic communicated (Suh 1999; Dennis et al. 2008). Thus, a serious disagreement with a colleague may not be best communicated through email, at least if there is a choice of face-to-face communication or telephone (Paul et al. 2004). However, for more routine issues an email may be much preferable since, despite its many drawbacks as a "cold medium," email does provide the significant advantage of permitting correspondents to choose their communication time.

Whole Team Communication

Particularly in large collaboration groups, there may be a high degree of task specialization, so much so that in some cases subgroups will fail to communicate with one another over long periods of time. In some instances this pattern may prove both efficient and effective, but it often presents hazards. In the first place, it is possible that one of the most common maladies of collaboration will strike: "false consensus," the illusion of consensus when in fact there is no consensus, just a failure to voice dissent (for an overview, see Marks and Miller 1987).

While the problem of false consensus is quite common in research collaborations (Hampton and Parker 2011; Cheruvelil et al. 2014), large, diffuse collaborations are especially prone. Second, and often even more problematic, it is easy in large, poorly communicating collaborations for team members to partition themselves into subgroups as mutually interested coalition partners. This can have disastrous consequences, such as mutual distrust among coalitions. The best way to avoid a partitioning into dysfunctional coalitions is to ensure that on at least some occasions there is whole-team communication. Certainly it is neither practical nor useful to have whole-team communication for all occasions and all activities, especially in the case of large research teams. The "art" is in knowing when all-team communications are called for and when they will just be a nuisance and a productivity killer. In our view, all-team communications tend to work best (1) at the very outset of the collaboration, when the team is organized and team members are getting acquainted; (2) when there are stakes or risks are high and clearly shared by all team members; and (3), when the collaboration is undergoing a major change, such as a new research team director, a new set of collaborators, or a change in the team's objectives (e.g., when transitioning from a sole focus on research and publication to application or commercialization).

High Level of Trust among Collaborators

In previous sections of the book, and in the Aggregate Model, we have discussed the role of trust in research collaborations, but it is worth returning to as an element of Consultative Collaboration. In collaborations of long standing, trust is often "built in" owing to previous shared experience. But in newer collaborations, or collaborations adding new team members, trust is a crucial goal. How to achieve trust? The easiest way is simply to abide by the other characteristics of Consultative Collaboration (e.g., occasional all-team communication, openness, and transparency). Announcing at the outset the collaboration management principles to be followed is quite useful in new collaborations and provides a standard for behavior, one that can be examined and, if followed, will almost certainly engender trust. The basic principles of Consultative Collaboration are designed to secure trust. If individuals feel that their ideas will be considered on their merits, not on the basis of reputation, status, or authorities, trust ensues. If communication is effective and open and if dissent is tolerated and diverse views welcomed, trust generally follows. If the collaboration recognizes different goals, values, and career needs, trust occurs more easily. In sum, trust is as much of a by-product of Consultative Collaboration Management as it is an attribute.

Disjunction of Ideas from the Individual

One of the more difficult admonitions from Consultative Collaboration Management is the separation of assessments of ideas from assessments of the individual proffering the idea. Group behavior researchers have long known that the quality of ideas is not well predicted by status or authority and that sometimes persons of relatively low status have good ideas rejected simply because of the source of the ideas (Berger et al. 1972; Driskell and Salas 1991; Kilduff et al. 2016). Several small-group organizational techniques have been developed to cope with the problems related to the ad hominum assessment of ideas (see Levi 2016 for an overview), most of them involving some sort of rules for anonymity of the source of ideas (Rohrbaugh 1981; McGrath and Hollingshead 1994; Hsu and Sandford 2007). Interestingly, many applications have been used in medical practice as a means to give voice to the ideas of nurses, which are too often overlooked by physicians (e.g., McKenna 1994; Keeney et al. 2006; Zega et al. 2014). One can easily substitute "graduate student" for "nurse" and "principal investigator" for "physician"; the same tendency to diminish the ideas of lower status colleagues exists in research teams.

In most cases, research team members do not need to become skilled in small group decision techniques to ensure that the ideas of lower status collaborators are given a fair hearing. Common sense goes a long way: asking collaborators their opinions rather than waiting for them to volunteer them, giving due consideration to ideas regardless of source, and, perhaps most important, simply being sensitive to the sometimes counterproductive tendency of most persons, not just researchers or team leaders, to color views about ideas offered with views about the person offering them.

Agreeing to Disagree, Openly

Also from the "playbook" of small group research is the problem of suppressing dissent and the related phenomenon of false consensus (Nemeth 1986; Tollefsen 2006), a phenomenon discussed in connection with all team communication. Classic studies of decision-making (Janis 1982; Maass and Clark 1984) tell us that bad things, sometimes tragic things (Smith 1985; Vaughan 1997; Ahlstrom and Wang 2009), can occur when legitimate disagreements are not brought to light, either due to an intolerant management structure or, just as problematic, a wish to avoid upsetting a false consensus. It is easy enough to see how this happens, even in a generally well functioning collaborative team. When most people seem to be in agreement, it is often very difficult for those who are in the minority, or who at least perceive themselves to be in the minority, to shatter what appears to be a happy consensus. If the person in the minority is a graduate student or postdoc whose job and career prospects are in large measure controlled by one or more of the team members on the other side of the disagreement, voicing dissent may well seem rash. Indeed, this is a rational view in such circumstances. Making a point, even a significant one, about a research problem pales in comparison to issues, real or perceived, of job security. Thus, it is unreasonable to expect that junior or career-dependent collaborators will be forthcoming with dissent. It is the responsibility of the PI and other senior collaborators actively to solicit the perspectives and opinions of more junior team members and, generally, to create a safe, open, and participative environment that will encourage response to such solicitations.

Revisiting Points of Agreement and Plans of Action

Open covenants openly arrived at, to quote Woodrow Wilson's comments about a very different collaboration, provide a sound basis for research collaboration. However, initial agreement does not always suffice for the life of the collaboration. Circumstances change, and agreements must be revisited. For example, there may be initial agreement about coauthor credit or order, but due to ordinary changes in participants' work agendas, the contributions to the collaboration may turn out to be different than originally anticipated. Similarly, sometimes collaborators find that one team member may be effectively deployed in ways different from those envisioned. Careers change, people take new jobs, take on new administrative responsibilities, are awarded new grants, take childbirth leave, gain or are denied tenure, get sick—in short, life happens and requires that we revisit collaboration agreements. How often should this occur? There can be no hard and fast rule, but Consultative Collaboration Management requires reconsideration when important events occur among the team members, and, for good measure, it is not a bad idea to set reconsideration and review benchmarks into initial collaboration management plans. A variety of techniques have been developed for team planning (for an overview, see Corey et al. 2014), and many of these, such as nominal group technique (Delbecq et al. 1975; McMillan et al. 2014), are simple and easily applied to the collaborative activities of research teams.

Recognition of Diverse Values and Objectives

As we see from the preceding chapters, collaborators often have very different values and objectives according to their respective career stages. However, not all differences on values and objectives flow from career stage. In addition to such factors as needing tenure or requiring a publication to obtain a new job, collaborators sometimes differ simply because they have different life and personal values, and these are reflected in the objectives they bring to research collaborations (Bozeman and Corley 2004). For example, some members of a research team may be focused almost entirely on developing the strongest possible scientific reputation and, relatedly, publishing in the highest-quality or most competitive journals. For others it may be at least as important to invest in mentoring the next generation of STEM researchers, and for still other team members commercial motivations may be paramount, leading them to focus on issues related to converting research into commercial technology applications. When we throw into the mix the sheer number of team members comprising some research collaborations as well as the inclusion of persons from diverse disciplines and fields, we can see that there is little likelihood of consensus on all significant collaboration decisions. Thus, the tendency to assume "everyone is on the same page" makes little difference when the "book" is written not only in different styles but is translated into different languages.

The Consultative Collaboration Management solution to dealing with the diversity of values and objectives motivations found in most large, multidisciplinary research teams is the same one that runs through most of our prescriptions: trust *and* verify, communicate about values and motivations rather than simply assuming that all are in agreement. This seems such a simple palliative for a problem that often seems intractable. Nonetheless, we find, time after time, that team members tend to assume that others' thinking is very much in line with their own. All too often that simply is not the case (Youtie and Bozeman 2016). It is easy enough to see why the particular delusion happens so often. Most researchers did not become scientists or engineers to manage anything. Some view management of processes and group relations as little more than the "tax" that must be paid in order to get to the work of research (McKelvey and Sekaran 1977; Hudson et al. 2002). Understandably, most wish to minimize the perceived tax, and one way of doing so, albeit a sometimes counterproductive way, is to simply assume (without any supporting evidence) that everyone is on the same page and then to invest minimally in team management and group processes. Moreover, there is evidence (Mosvick 1971; Lounsbury et al. 2012; Ryan 2014) that by their very nature most STEM researchers are not the most "touchy feely" human relations–focused people, though we have seen people at all points of the introvert-extrovert distribution. Making matters worse, at least from a research management perspective, research collaboration trends make it more difficult to engage in effective communication of values, motivations, and needs.

The increasing number of persons involved in collaborations (Lee, Walsh and Wang 2015) and the tendency to collaborate with persons in other disciplines, other institutions, and even other nations increases the diversity of values and objectives and, at the same time, makes recognition and accommodation all the more difficult (Jeffrey 2003). When we put a complex overlay of cultural differences into the collaborative mix (Pieterse et al. 2013), the potential for human relations problems increases accordingly. Conceptually, it is easy enough to understand the utility of consultative process and management, but we are not discounting the difficulty of taking the time and energy of actually doing so. We are simply saying that the investment usually pays off, and especially so in new collaborations or old ones with new team members.

Sensitivity to Differences in Status and Power

Even bearing in mind our admonition above to take into account differences in team members' values and motivations, and managing research to make productive uses of those differences rather than allowing them to threaten the collaboration, we advise giving particular attention to the special demands of teams composed of persons with very different status and power. Our results, regardless of type of data, reveal the same tendency: less experienced and less powerful researchers are the ones most often providing problems in research collaboration. Consultative Collaboration Management recognizes that those in lesser power positions, such as students, postdocs, and junior faculty may not always speak up and share their views about motives and values (Hara et al. 2003). Nor should they be expected to do so. Typically, the senior members of research teams have resources crucial to those who have little or no control of team resources. Thus, Collaborative Collaboration Management puts the responsibility on senior research members and directors for directly soliciting and listing to the views and concerns of (usually silent) subordinates and inexperienced team members. Naturally, we do not expect tyrants to heed this advice. However, most management problems related to insensitivity to power and status differences do not occur because of tyrannical research management but because research managers who are benevolent or at least innocuous do not focus sufficient attention on the needs of more junior team members. As mentioned above, Directive Collaboration Management and Assumptive Collaboration Management are by far the most common approaches, and thus many senior research investigators and managers have not participated in research teams that systematically attend to the views of the less powerful.

Conclusion: A Final Word

Our final word? That there is no final word. Research collaboration is too variegated and rapidly changing. Yet at the same time, treating each problem idiosyncratically is both impractical and unnecessary. Our work shows that one must be wary of cookbook approaches to research collaboration improvement, but it is possible to draw lessons and judiciously apply them. Here we suggested a Consultative Collaboration Management approach that we feel could forestall a great many problems research teams encounter. The approach is not formulaic but rather entails a general commitment to eliciting all collaborators' views honestly and validly and working with them to proceed in directions that, to the extent possible, fulfill the sometimes diverse objectives of team members. Consultative Collaboration Management is an approach that provides only general cues for improving research team effectiveness. By contrast, the Aggregate Collaboration Effectiveness Model put forth in chapter 4 and deployed in this chapter highlights specific factors affecting research collaboration outcomes. Taken together, the lessons from the two different prescriptive approaches should apply well to the vast majority of research teams.

At the risk of being too simplistic, if we had to sum up the chief lesson we learned from this research, it comes down to the need for better communication and more frequent and more systematic attention to the social processes underpinning research teams' collaborations. Understandably, collaborators, especially experienced ones, develop de facto strategies based on their own experience, and they often reflexively execute those strategies. Ad hoc strategies or accumulated routines may not always be a great idea. We find that many researchers continue to commit the same sins, sometimes because of their early socialization with the first collaborators and sometimes because they make assumptions about others' norms and perceptions, assumptions that often turn out to be wrong. Usually there is nothing malicious in intent, but a lack of malice is not the point. With more communication and attention to social dynamics and, particularly, by breaking the frame of assumed agreement and consent, research collaboration outcomes can be improved.

The terms *intentionality* and (please forgive the cliché) *mindfulness* come to us as watchwords for effective team research collaboration. The Consultative Collaboration Management approach entails prescriptions that seem to us remarkably simple and straightforward. *However, most collaborators do not follow this approach.* Our survey results point out that the majority of the 640 responding scholars do not have explicit discussions about collaboration strategies and management. As we saw in chapter 4, only 40 percent of survey respondents have explicit discussions about co-authorship credit. Among those who do have such explicit discussions, less than one-third conduct such discussions in the early stages of the research. Explicit discussions about collaboration management and, especially, about crediting are less common with research involving younger researchers (graduate students, postdocs, assistant professors) (Youtie and Bozeman 2014). Consultation and explicit discussions are shown to reduce problematic collaborations, but the work habits of most researchers do not embrace this approach. Why is this the case? One interview subject told us thus:

> I am having a problem right now. It is important to her, my collaborator, to decide early on who should be an author. I don't understand why she thinks it is important. In tenure and promotion you are expected to produce publications. But why is it worth so much effort on our part? Sometimes other people are invested in this. What is the process? They see outcomes-based issues. I am so divested in this. . . . I don't want to put in any effort in this.

This lack of up-front communications harkens back to the discussion of personalities in chapter 5. Sometimes a lack of communication is owing to the "jackass factor," difficult researchers who do not much care about others' values. This book is not aimed at the problem of self-consciously exploitative individuals engaging jn unrepentantly bad behavior. However, in many cases the lack of consultation and limited or ineffective communication occurs with generally well-meaning research team members who in general have good research values and good will (or at least a lack of ill will) toward their collaborators. As we have seen, failure to consult sometimes is explained by the false assumption of consensus, as per the Assumptive Collaboration Management Model. But in other cases research team members are, for any of a variety of reasons too passive. Passive team members may quietly assent to unsatisfactory author orders or unfair crediting. Passive researchers may convince themselves that they are being generous in including noncontributing authors when in fact they are helping noncontributors to build undeserved reputations and social capital. Passive researchers, not sufficiently attuned to the educational function of collaboration, may avoid training or socializing graduate students in the ethics of collaboration and the appropriate expectations of research team members.

Fortunately, most research team members are well aware of the need to pass along to new generations of researchers the norms, tacit knowledge, and ethical standards pertaining to research collaboration and management of research teams. We hope this book will help them in this task.

Data Sources, Methods, and Research Procedures

We have developed a considerable amount of original data for this book, but we also make extensive use of previous work on research collaboration and work from the general literature on the topic, as well as work from the set of colleagues with whom we have interacted for more than fifteen years in studies of research collaboration, including a project known as the Scientific and Technical Human Capital Project. Our book relies on three distinctive original data sets: (1) survey questionnaire-based; (2) interview-based; and (3) based on anonymous posts on a Website. We discuss each in turn.

Survey

We conducted a Web survey of nonmedical academic researchers in science and technology disciplines in US doctoral research universities (Carnegie Doctoral/Research Universities—High). A sampling frame of science and technology fields was developed using NSF's categories in its Survey of Earned Doctorates. Health sciences was excluded (because of its medical orientation), while economics was added to incorporate social science practices into the survey. The resulting frame was based on fourteen disciplines in biology, chemistry, computer science, mathematics, engineering, and economics. The sampling frame called for one male and one female faculty member from each randomly selected department at a given university because qualitative interviews suggested that gender would be a significant factor; in the event that no female faculty members were affiliated with the department, two male researchers were selected. The target sampling frame, which assumed a 50 percent response rate, resulted in 2,996 faculty and another 216 post-docs[1]. We were able to collect contact information for 2,574 individuals in the sampling frame; of these 2,189 were of sufficient quality as indicated by an electronic mail verification software program. Pilot surveys performed in April and May of 2012 used 400 of these, leaving 1,789 for the final survey. Our original idea had been to survey coauthors of a published journal article; we designed a survey around this initial approach and piloted it in April 2011 with 300 respondents; this pilot yielded a low response rate of 10 percent. We followed up with 30 nonrespondents to the pilot survey, learning that reluctance to participate in a survey about explicitly named coauthors was a problem. This finding led us to redesign the sampling frame to focus on randomly selected male and female faculty members in the aforementioned universities and to redesign the questionnaire to refer to "the last coauthored research publication you had accepted for publication in a scientific or professional journal" without naming the publication but emphasizing the most recent one so that the experience would be fresh in the respondent's mind. We also shortened the Web questionnaire so that it focused on (1) experiences associated with the most recent coauthored research publication such as the number of coauthors, number of male/female coauthors, location of coauthors, sector of coauthors, coauthorship position of the respondent, year accepted for publication, motivations for collaboration, relationship to coauthors before the paper, primary basis for coauthor ordering, decision-making process, research activity by coauthor position, views about coauthor experience; (2) career-long experiences with collaboration, such as the percentage of coauthored publications overall and with students, and extent of experience with potential negative behaviors; and (3) background information such as year of doctoral degree, year of tenure, rank/title, male/female, racial/ethnic identity, birth year, and number of doctoral and postdoctoral students supervised.

Six waves of survey invitations and reminders were sent in October and November of 2012. In all, we received 641 completed or mostly completed online questionnaires, for a 36 percent response rate. Respondents were very similar to the population in terms of gender, rank, and departmental discipline. Among nonrespondents, 1 percent were not at their office location, while another 5 percent explicitly opted out of participation. Given that we oversampled females and certain departments, we re-weighted the results to reflect the population distribution as indicated in the NSF Survey of Doctorate Recipients 2006 (the most recently available survey at the time the initial analyses were performed[2]). A breakdown of survey responses by survey question can be found at the end of this appendix (appendix table 1). Throughout this book we use simple frequency breakdowns or cross tabulations from the survey, although in another work, we have used regression models to characterize, for example, relationships between decision-making and likelihood of problems (Youtie and Bozeman 2014).

Interviews

A second approach involved in-depth interviews with sixty STEM researchers, all professors in research-intensive universities. In addition, interviews with economists were also conducted to serve as a comparison group for extending the STEM results to other disciplines. The STEM disciplines were biology, computer science, mathematics, physics, earth and atmospheric science, chemistry, chemical engineering, economics, civil engineering, electrical engineering, mechanical engineering, materials engineering/materials science, industrial/manufacturing engineering, and biomedical engineering.

In-person interviews were conducted with faculty at geographically distributed universities in Arizona, California, Georgia, South Carolina, Massachusetts, and Illinois. Additional telephone interviews were conducted with faculty at universities in Ohio, Michigan, Illinois, Texas, and Pennsylvania. Of the interviews, 32 were conducted in person, 27 by telephone, and one conducted via Skype because the interview subject was working in Japan. Interview subjects were selected at random based on their affiliation with schools and departments in the 14 STEM disciplines plus economics. Only assistant, associate, and full professors were selected to ensure sufficient range of research collaboration for discussion. Of the interviews, 33 were with full professors, 16 were with associate professors, and 11 were with assistant professors. We interviewed more senior faculty than junior faculty, chiefly because we wished to have a diverse array of experiences, and researchers with only two or three years' collaborating experience typically do not have as much to draw upon. Although the majority of the interview subjects were men, 17 of the 60 interview subjects were women. The interviews were held in the spring, summer, and autumn of 2012 and the spring of 2013, typically in the offices of the faculty being interviewed.

Each interview lasted from thirty minutes to an hour or more. Interviews were not taped, rather notes were taken either by hand or computer. Interview questions covered topics about the faculty's research characteristics, nature of research collaborations, decision-making processes for determining authorship and author order, role of policies such as promotion and tenure, and stories about best and worst research collaborations. The interview protocol allowed for flexibility to adapt to each interview subject's situation. For example, some interview subjects did not initially report ever having any contributorship problems. The interviewer was able to use prompts such as asking about a colleague's problematic experiences or "urban legends" in the field to obtain such information; sometimes these accounts from other colleagues would prompt interview subjects to recollect problematic collaborations of their own. While the focus was the same in every case—learning about bad and good research collaboration experiences and identifying factors related to effectiveness— the interviews were sufficiently expansive to allow long digressions when these were of interest. Thus, if one interviewee had great interest and knowledge in gender dynamics, we focused more on this topic; if another had extensive experience collaborating with industry, we spent more time talking about this.

The results of these interviews were collated by the project team. Each interview was read in its entirety. For the purpose of this paper, comments with relevance to resolution of problematic research collaborations were extracted and analyzed. Two graduate students prepared and organized the text from all the interviews. We then reviewed their work, looking for common as well as unexpected themes and representative quotes.

Anonymous Web Posts

While our interviews were remarkably helpful, we had some concerns that respondents may have been a bit less anxious to talk face-to-face about the worst of their collaboration experiences. Thus, we rely on a third method, anonymous Web posts. When we administered our structured questionnaire, we at the same time asked our questionnaire respondents to consider posting comments on a different and fully anonymous Website, telling us about of their best and worse collaboration experiences. Initially we asked for best and worst for symmetry's sake (since we found no one had difficulty talking in the interviews about their best collaborations), but we learned some new things in the Web posts about the best collaborations. In total, 93 survey respondents posted 161 Website posts, of which 44 percent were about positive experiences and 56 percent about negative experiences. The posts have been coded in a similar manner to the interview text, and especially poignant experiences were highlighted for use throughout this book.

National Study of Research Collaboration Survey Results

(Weighted results to reflect the population distribution by gender and field as indicated in the NSF Survey of Doctorate Recipients 2006)

Question: "Regarding [the] most recent coauthored research publication, how many coauthors were listed in the publication (including yourself)?"

2 authors	26.3%
3 authors	17.2%
4 authors	14.6%
5 authors	12.5%
6–7 authors	15.3%
8–10 authors	7.6%
11–36 authors	4.9%
57–2,000 authors	1.6%
	100.0%
Observations	635

Question: "Were you the first listed author in the publication?"

	Valid %
No	73.6%
Yes	26.4%
	100.0%
Observations	641

Question: "Regarding this recent coauthored research publication, in what year was it accepted for publication?"

2012	71%
2011	16%
2010	6%
2009	3%
1998–2008	5%
	100%
Observations	626

Question: "Researchers may have many different motivations for collaborating, including (but not limited to) the ones listed below. Regarding the factors that led you to collaborate on this most recent coauthored research paper, how important, if at all, were the following?"

	(percentage of respondents shown below)										
	Not important at all									Extremely important	
	1	2	3	4	5	6	7	8	9	10	# Observations
Helping me to obtain tenure or promotion	53.8	2.4	3.1	4.7	12.0	3.3	3.1	6.3	3.1	3.5	534
Helping a coauthor's career	21.2	2.8	4.1	3.6	21.9	2.4	6.1	10.1	9.9	5.7	612
Working with persons highly fluent in my native language	74.1	5.8	4.9	1.3	6.5	1.9	1.1	0.9	1.5	2.0	547
Providing research training for coauthoring students or postdocs	20.5	1.3	2.4	1.6	14.7	3.9	6.0	7.5	12.5	6.4	602
Having fun working with researchers I like on a personal basis	10.0	1.7	3.5	2.2	18.1	2.6	9.3	12.5	15.2	6.9	621
Increasing my own research productivity	3.2	1.7	2.0	1.8	12.7	1.9	8.8	12.6	21.8	7.9	630
Working with researchers whose skills and knowledge complement mine	2.1	0.5	0.7	0.8	6.0	0.7	4.7	11.5	20.4	8.8	627
Working with persons I can depend on to complete work on time	5.1	2.1	1.4	1.8	20.5	3.7	6.7	15.9	21.3	7.3	623
Working with collaborator(s) whom I felt would be fair in coauthor crediting or order of authorship	10.1	5.5	6.3	2.3	22.6	2.5	5.0	13.1	15.0	6.3	611

Question: "How would you describe your relationship to your coauthor(s) before the collaborative paper?"

	Percentage	Observations
My student, former student/postdoc	38.6%	247
Same academic department	31.4%	201
Met at conference	26.3%	168
Through a mutual colleague	22.1%	142
Same university, different departments	19.8%	127
Met in school	13.0%	83
Was my thesis advisor	5.6%	36
Other (e.g., wrote to one another)	9.2%	59

Question: "Did you and your coauthor(s) ever have explicit discussions about coauthoring credit?"

We had explicit discussions	40%
We had no explicit discussion, these issues were more or less assumed	60%
Observations	641

Question (if explicit discussions): "When did the discussions about coauthorship occur? (Discussions may have occurred on several occasions, so please check all that apply.)"

Discussions about coauthor credit occurred even before the research commenced	16%
Discussions occurred as soon as the research commenced	15%
Discussions occurred during the course of the research	67%
Discussions occurred after the research was concluded	50%
Observations	256

Question (if no explicit discussions): "How was the decision made about whom to include and not include as a coauthor? Please check all that apply."

	Observations	Valid percent
Coauthor order is alphabetic	118	18.5%
Order is random	3	0.4%
Coauthor order is based on decisions made by one or more coauthors	63	10.0%
Order reflects one or more of the coauthors assessment of the importance of each coauthor's research contribution	330	51.9%
The most senior or highly reputed authors are listed before the less senior ones	9	1.4%
The most senior or highly reputed authors are listed after the less senior ones	91	14.3%
Order is based on agreements or procedures adopted from coauthors earlier collaborative research	17	2.7%
I do not know how we arrived at this rationale for the ordering of coauthors	<u>5</u>	<u>0.9%</u>
Total	637	100.0%

Question: "Please tell us the gender of the coauthors (including yourself).... At the time the research paper was first submitted to the journal ultimately accepting it for publications, where were your coauthors located?"

Number of coauthors	Male	Female	Affiliated with your university (not including you)	In a different university	In a private firm or industry
0	2.3%	37.3%	23.4%	19.9%	81.8%
1	14.4%	29.9%	27.1%	31.1%	10.3%
2	24.1%	16.5%	20.7%	19.0%	5.6%
3	22.9%	8.8%	10.9%	11.1%	1.3%
4–5	22.5%	4.6%	11.6%	9.0%	0.7%
6 or more	<u>13.8%</u>	<u>2.9%</u>	<u>6.3%</u>	<u>9.9%</u>	<u>0.3%</u>
	100.0%	100.0%	100.0%	100.0%	100.0%
Observations	608	607	568	506	306

Question: "Below is a list of activities that could be part of any research collaboration. Referring again to the same most recent coauthored publication, please indicate below whether coauthors were engaged in the respective activities."

Task	Lead author	Coauthor	Person not listed as a coauthor
Research question	42.2%	26.6%	31.2%
Data	24.7%	39.0%	36.3%
Data analysis	25.2%	40.2%	34.6%
Writing text	27.6%	41.1%	31.3%
Literature review	26.8%	41.0%	32.2%
Grant/contract funds	29.0%	31.7%	39.3%
Administration	22.4%	21.7%	55.9%
Observations	541		

Question: "Below is a series of statements that may or may not relate to your views about your coauthoring experience regarding this most recent publication. Please indicate the extent to which you agree or disagreement with each statement." (strongly agree = 1, strongly disagree = 10, number of observations = 629, percentages sum across the rows to 100%)."

	Percentage of Respondents									
	Strongly agree								Strongly disagree	
	1	2	3	4	5	6	7	8	9	10
I consider this publication to be one of my best in terms of scientific quality	10.5	12.8	17.6	10.2	5.7	27.9	8.0	5.2	0.8	1.3
One or more of the coauthors was strongly motivated by the commercial possibilities of the research	1.8	1.6	1.9	2.4	1.5	9.2	1.9	7.4	7.9	64.4

Continued on next page

	Percentage of Respondents (*continued*)									
	Strongly agree							Strongly disagree		
	1	2	3	4	5	6	7	8	9	10
One or more of the coauthors was strongly motivated by the potential of the research to help solve social problems	9.0	5.7	7.2	5.7	6.5	10.5	2.5	4.3	5.4	43.2
All things considered, I feel my contribution was greater than my coauthors	8.6	4.2	5.2	4.4	3.6	29.6	6.2	7.5	7.9	22.8
There is at least one person who deserved coauthor credit but did not receive it	2.4	1.6	2.9	1.6	1.0	5.9	1.9	2.4	4.2	76.1
There is at least one person who did not deserve coauthor credit but received it	1.0	0.3	0.2	0.2	0.3	3.6	0.2	1.9	3.8	88.5
There was significant gender-based conflict among the coauthors	1.7	0.2	0.3	0.3	0.2	3.4	0.0	0.7	2.5	90.7
Observations	629									

Question: "For your entire research career, approximately what percentage of your published research have been single authored (i.e., no coauthors)?"

% research	% respondents
0%	33.5%
1–4%	15.5%
5–9%	13.8%
10–24%	19.4%
025%–50%	11.7%
51%–99%	5.6%
100%	0.5%
Total	100.0%
Observations	522

Question: "For your entire research career, approximately what percentage of your coauthored papers have included students as coauthors?"

% research	% respondents
0%	5.1%
1–15%	10.8%
16–30%	10.7%
31–50%	19.0%
51–80%	20.2%
81–99%	18.2%
100%	15.5%
Total	99.5%
Observations	519

Question: "Below is a list of negative behaviors that sometimes occur in research collaboration. Have you at any point had such experiences in your collaborations?"

	Has never happened	Happened 1–3 times	Happened more than 3 times	Observations
A coauthor did not finish agreed upon research-related activities	31.4%	48.9%	19.7%	625
Coauthor claimed lead authorship when it was not deserved	72.5%	24.8%	2.7%	627
A person listed as a coauthor made no contribution at all to the research	60.0%	31.5%	8.5%	626
Coauthorship credit was denied to someone who deserved to be a coauthor	83.7%	0.2%	16.1%	627

Question: "In what year did you receive your doctoral degree?"

Year	% respondents
1958–1980	22.0%
1981–1990	24.3%
1991–2000	29.6%
2001–2012	<u>24.1%</u>
Total	100.0%
Observations	624

Question: "What is your current rank and title?"

Rank/Title	% respondents
Professor	58.0%
Associate professor	20.4%
Assistant professor	17.4%
Postdoc	3.5%
Other	<u>0.7%</u>
Total	100.0%
Observations	641

Question (unweighted results): "Are you female or male?"

Female	51.6%
Male	48.4%
Observations	641

Question: "What is your racial, ethnic identity?"

Identity	% respondents
White	78.7%
Asian	12.1%
African/Black	1.0%
Hispanic	4.5%
Native American	0.3%
Other	0.5%
Decline to say	2.9%
Observations	641

Question: "During the past five years, what is the total number (if any) of doctoral students for whom you have been their primary dissertation supervisor?"

# students	% respondents
0	13.5%
1	10.6%
2	14.7%
3	12.6%
4	10.6%
5	9.1%
6–7	10.0%
8–9	4.6%
10	6.7%
11+	7.6%
Total	100.0%
Observations	610

Question: "During the past five years, what is the total number (if any) of postdoctoral students for whom you have been their primary dissertation supervisor?"

# students	% respondents
0	44%
1	19%
2	14%
3	8%
4	5%
5	4%
6–9	5%
10+	3%
Total	100%
Observations	612

APPENDIX 2

Propositional Literature Table

Collaboration topic	In-text citation	Full citation	Relevant findings
		Definition of Research Collaboration	
Level of Analysis	Youtie and Bozeman 2014	Youtie, J., and Bozeman, B. (2014). Social dynamics of research collaboration: Norms, practices, and ethical issues in determining co-authorship rights. *Scientometrics*, 101(2), 953–962.	Research collaboration effectiveness can be measured at the organizational level but collaborations are formed by individuals.
Level of Analysis	Burt 2000	Burt, R. S. (2000). The network structure of social capital. *Research in Organizational Behavior*, 22, 345–423.	Social network analysis techniques may be helpful for future study of research collaborations.
Level of Analysis	Clark and Mills 2011	Clark, M. S., and Mills, J. R. (2011). A theory of communal (and exchange) relationships. *Handbook of Theories of Social Psychology*. Sage Publications, Los Angeles, CA, 232–250.	Social exchange theory may also provide some insights into research collaboration effectiveness.
Defining Collaboration	Katz and Martin 1997	Katz, J. S., and Martin, B. R. (1997). What is research collaboration? *Research Policy*, 26(1), 1–18.	Most research collaborations are defined by coauthorship of peer-reviewed publications, but this definition has limitations.
Defining Collaboration	Shrum et al. 2001	Shrum, W., Chompalov, I., and Genuth, J. (2001). Trust, conflict and performance in scientific collaborations. *Social Studies of Science*, 31(5), 681–730.	Research collaborations are better described by pooling of human capital for the purpose of producing knowledge.

Continued on next page

Collaboration topic	In-text citation	Full citation	Relevant findings
Defining Collaboration	Youtie and Bozeman 2014	Youtie, J., and Bozeman, B. (2014). Social dynamics of research collaboration: norms, practices, and ethical issues in determining coauthorship rights. *Scientometrics*, 101(2), 953–962.	Many research collaborations produce nothing more than future opportunities for collaboration. A research collaboration is not defined by the immediate outcomes.
Defining Collaboration	Flipse et al. 2014	Flipse, S. M., van der Sanden, M. A., and Osseweijer, P. (2014). Setting up spaces for collaboration in industry between researchers from the natural and social sciences. *Science and Engineering Ethics*, 20(1), 7–22.	Research collaborations are not defined by specific outcomes, since they often complicated by intellectual property issues, especially when academic-industry partnerships are involved.
Defining Collaboration	Bozeman et al. 2013	Bozeman, B., Fay, D., and Slade, C. P. (2013). Research collaboration in universities and academic entrepreneurship: The-state-of-the-art. *Journal of Technology Transfer*, 38(1), 1–67.	Being a patron, meaning making financial contribution to a research project, does not make the patron a research collaborator.
Collaboration as Human Capital	Ponomariov and Boardman 2010	Ponomariov, B., and Boardman, P. C. (2008). The effect of informal industry contacts on the time university scientists allocate to collaborative research with industry. *Journal of Technology Transfer*, 33(3), 301–313.	Scientific and technical human capital is a unique form of human capital that involves the results of researchers' professional ties, technical skills, and shared efforts.
Collaboration as Human Capital	Stokes and Hartley 1989	Stokes, T. D. and Hartley, J. A. (1989). Coauthorship, social structure and influence within specialties. *Social Studies of Science*, 19(1), 101–125.	Joint laboratory work involves research, but it does not necessarily rise to the level of a research collaboration.

Collaboration topic	In-text citation	Full citation	Relevant findings
Collaboration as Human Capital	Muller 2014	Müller, R. R. (2014). Postdoctoral life scientists and supervision work in the contemporary university: A case study of changes in the cultural norms of science. *Minerva: A Review of Science, Learning and Policy*, 52(3), 329–349.	Advising a student or providing mentorship on research does not necessarily rise to the level of research collaboration.
Research Collaboration as Co-Authorship	Heffner 1981	Heffner, A. G. (1981). Funded research, multiple authorship, and subauthorship collaboration in four disciplines. *Scientometrics*, 3(1), 5–12.	Research productivity is typically measured by peer-reviewed journal authorship.
Research Collaboration as Co-Authorship	Vinkler 1993	Vinkler, P. (1993). Research contribution, authorship and team cooperativeness. *Scientometrics*, 26(1), 213–230.	Research productivity is typically measured by peer-reviewed journal authorship.
Research Collaboration as Co-Authorship	Melin and Persson 1996	Melin, G., and Persson, O. (1996). Studying research collaboration using coauthorships. *Scientometrics*, 36(3), 363–377.	Research productivity is typically measured by peer-reviewed journal authorship.
Research Collaboration as Co-Authorship	Wagner 2005	Wagner, C. S. (2005). Six case studies of international collaboration in science. *Scientometrics*, 62(1), 3–26.	Research productivity is typically measured by peer-reviewed journal authorship.
Research Collaboration as Co-Authorship	Heinze and Bauer 2007	Heinze, T., and Bauer, G. (2007). Characterizing creative scientists in nano-S&T: Productivity, multidisciplinarity, and network brokerage in a longitudinal perspective. *Scientometrics*, 70(3), 811–830.	Publication and citation scores are typical measures of scientific productivity, but novel ideas like science awards and nominations are gaining traction.
Research Collaboration as Co-Authorship	Mattsson et al. 2008	Mattsson, P., Laget, P., Nilsson, A. and Sundberg, C. (2008). Intra-EU vs. extra-EU scientific co-publication patterns in EU. *Scientometrics*, 75(3), 555–574.	The use of publications and citation scores is an international standard. It allows for observation of changes, such as increasing coauthored papers.

Continued on next page

Collaboration topic	In-text citation	Full citation	Relevant findings
Research Collaboration as Co-Authorship	Mayrose and Freilich 2015	Mayrose, I., and Freilich, S. (2015). The interplay between scientific overlap and cooperation and the resulting gain in coauthorship interactions. *Plos ONE*, 10(9), 1–10.	Evaluating coauthorship patterns can provide insights into scientific interactions that may explain productive science.
New Measures of Research Collaboration	Melin 2000	Melin, G. (2000). Pragmatism and self-organization: Research collaboration on the individual level. *Research Policy*, 29(1), 31–40.	New and interesting approaches are evolving to measure individual motivations for research and research collaboration.
New Measures of Research Collaboration	Bozeman and Corley 2004	Bozeman, B., and Corley, E. (2004). Scientists' collaboration strategies: Implications for scientific and technical human capital. *Research Policy*, 33(4), 599–616.	This study recognizes the importance of researcher self-reported collaboration choices and strategies.
New Measures of Research Collaboration	Bozeman and Gaughan 2011	Bozeman, B., and Gaughan, M. (2011). How do men and women differ in research collaborations? An analysis of the collaboration motives and strategies of academic researchers. *Research Policy*, 40(10), 1393–1402.	Researcher self-reported collaboration choices and strategies collected through a survey can effectively explain interesting and important collaboration patterns.
New Measures of Research Collaboration	Youtie and Bozeman 2014	Youtie, J., and Bozeman, B. (2014). Social dynamics of research collaboration: Norms, practices, and ethical issues in determining coauthorship rights. *Scientometrics*, 101(2), 953–962.	Qualitative research methods including semi-structured interviews provide insights into collaboration patterns, especially across disciplines and in particular decisions to coauthor or not.

Collaboration topic	In-text citation	Full citation	Relevant findings
New Measures of Research Collaboration	Jeong et al. 2011	Jeong, S., Choi, J. Y., and Kim, J. (2011). The determinants of research collaboration modes: exploring the effects of research and researcher characteristics on co-authorship. *Scientometrics*, 89, 3, 967–983.	Researchers are accountable for selecting their "collaboration modes," such as sole research, internal collaboration, domestic collaboration, and international collaboration, where they make strategic decisions about who to collaborate with and the trade-offs associated with collaboration.
New Measures of Research Collaboration	Laudel 2001	Laudel, G. (2001). Collaboration, creativity and rewards: Why and how scientists collaborate. *International Journal of Technology Management*, 22, 762–81.	Collaborative research occurs both within and between research groups with differing rewards for associated scientists.
Assessing Research Collaborations with Systematic Thinking			
Exploring Research Collaborations at the System Level	Hou et al. 2008	Hou, H. Y., Kretschmer, H., Liu, Z. Y. (2008). The structure of scientific collaboration networks in Scientometrics. *Scientometrics*, 75(2), 189–202.	Scientometrics is evolving by use of bibliographic data to explore collaboration at many levels of analysis.
Exploring Research Collaborations at the System Level	Bozeman et al. 2013	Bozeman, B., Fay, D., and Slade, C. P. (2013). Research collaboration in universities and academic entrepreneurship: the-state-of-the-art. *Journal of Technology Transfer*, 38(1), 1–67.	Research collaborations should be studied at many levels of analysis. including the attributes of individual collaborators, and the collaborative process at the organization level.
Exploring Research Collaborations at the System Level	Vasileiadou 2012	Vasileiadou, E. (2012). Research teams as complex systems: Implications for knowledge management. *Knowledge Management of Research Practice*, 10(2), 118–127.	Research teams are complex systems that emerge from individuals working in groups, with all associated local and global dynamics.

Continued on next page

Collaboration topic	In-text citation	Full citation	Relevant findings
Exploring Research Collaborations at the System Level	Hessels 2013	Hessels, L. K. (2013). Coordination in the science system: Theoretical framework and a case study of an intermediary organization. *Minerva: A Review of Science, Learning and Policy*, 51(3), 317–339.	Intermediary organizations, including those that provide funding, are affecting the coordination and outcomes of research collaborations.
Exploring Research Collaborations at the System Level	Bozeman et al. 2015	Bozeman, B., Rimes, H., and Youtie, J. (2015). The evolving state-of-the-art in technology transfer research: Revisiting the contingent effectiveness model. *Research Policy*, 44(1), 34–49.	Public values beyond individual economic interests can explain effective research collaboration. This is the contingent effectiveness model of technology transfer.
Exploring Research Collaborations at the System Level	Shrum et al. 2007	Shrum, W., Genuth, J., and Chompalov, I. (2007). *Structures of Scientific Collaboration*. MIT Press, Cambridge, MA.	The work of individual researchers is part of a general trend toward more fluid, flexible, and temporary organizational arrangements for research collaborations.
Exploring Research Collaborations at the System Level	Ulnicane 2015	Ulnicane, I. (2015). Why do international research collaborations last? Virtuous circle of feedback loops, continuity and renewal. *Science and Public Policy* (SPP), 42(4).	There is a general trend toward more fluid, flexible, and temporary organizational arrangements, such as teams, for research collaborations.
Exploring Research Collaborations at the System Level	Bozeman et al 2013	Bozeman, B., Fay, D., and Slade, C. P. (2013). Research collaboration in universities and academic entrepreneurship: The-state-of-the-art. *Journal of Technology Transfer*, 38(1), 1–67.	Attention should be given university researchers' collaborations with researchers in other sectors, including industry.
Exploring Research Collaborations at the System Level	Bozeman and Rogers 2002	Bozeman, B. and Rogers, J. (2002). A churn model of knowledge value: Internet researchers as a knowledge value collective. *Research Policy*, 31(4), 769–794.	Research collaboration occurs on the Internet as a new mode and system of knowledge creation.

Collaboration topic	In-text citation	Full citation	Relevant findings
Systems of Social Exchange	Emerson et al. 1976	Emerson, R. M. (1976). Social exchange theory. *Annual Review of Sociology*, 335–362.	Social exchange theory is useful in determining how research collaborators form and work in teams.
Systems of Social Exchange	Mitchell et al. 2015	Mitchell, G. E., O'Leary, R., and Gerard, C. (2015). collaboration and performance: Perspectives from public managers and NGO leaders. *Public Performance and Management Review*, 38(4), 684–716.	Forming research teams is both costly and stressful and that can result in less than optimal research collaboration results.
Systems of Social Exchange	Li et al. 2013	Li, E. Y., Liao, C. H., and Yen, H. R. (2013). Co-authorship networks and research impact: A social capital perspective. *Research Policy*, 42(9), 1515–1530.	Systems of researchers and outcomes of research evolve based on social relationships and issues such as trust and distrust between collaborators.
Cross-Sector Research Collaborations	Gray 2011	Gray, D. O. (2011). Cross-sector research collaboration in the USA: A national innovation system perspective. *Science and Public Policy*, 38(2), 123–133.	Cross-sector research collaborations, including those involving multiple levels of state, local and federal funding, are an evolving system of science, technology and innovation.
Cross-Sector Research Collaborations	Clark 2011	Clark, B. Y. (2011). Influences and conflicts of federal policies in academic–industrial scientific collaboration. *Journal of Technology Transfer*, 36(5), 514–545.	The federal government, primarily through funding, shapes systems of research collaborations.
Cross-Sector Research Collaborations	Hagstrom 1965	Hagstrom, W. O. (1965). *The Scientific Community*. Basic Books, New York, NY.	Discussion about the impact of federal funding and government priorities for research on systems of research collaboration has dated back several decades.

Continued on next page

Collaboration topic	In-text citation	Full citation	Relevant findings
		Norms and Decision-Making	
Research Collaboration Decision-Making Processes	Brockner and Wiesenfeld 1996	Brockner, J., and Wiesenfeld, B. M. (1996). An integrative framework for explaining reactions to decisions: Interactive effects of outcomes and procedures. *Psychological Bulletin*, 120(2), 189–208.	Decision-making processes determine outcomes of research collaborations.
Research Collaboration Decision-Making Processes	Katz and Martin, 1997	Katz, J. S., and Martin, B. R. (1997). What is research collaboration? *Research Policy*, 26(1), 1–18.	Decision-making processes and norms affect development of science in many ways.
Research Collaboration Decision-Making Processes	Melin 2000	Melin, G. (2000). Pragmatism and self-organization: Research collabora-tion on the individual level. *Research Policy*, 29(1), 31–40.	Decision-making processes and norms affect development of science in many ways.
Research Collaboration Decisions and Academic Careers	Heffner 1981	Heffner, A. G. (1981). Funded research, multiple authorship, and subauthorship collabo-ration in four disciplines. *Scientometrics*, 3(1), 5–12.	Research collaboration decisions continue to use authorship as a determining factor.
Research Collaboration Decisions and Academic Careers	Schut et al. 2014	Schut, M., van Paassen, A., Leeuwis, C., and Klerkx, L. (2014). Towards dynamic research config-urations: A framework for reflection on the contribution of research to policy and innovation processes. *Science and Public Policy*, 41(2), 207.	There are complex interpersonal dynamics, including competing claims for recognition, that shape the decision-making and contribution of researchers to policy and innovation processes.
Research Collaboration Decisions	Mayrose and Freilich 2015	Mayrose, I., and Freilich, S. (2015). The interplay between scientific overlap and cooperation and the resulting gain in co-authorship interactions. Plos ONE, 10(9), 1–10.	Shared research in-terests and agendas weigh heavily on researchers' deci-sions to collaborate.

Collaboration topic	In-text citation	Full citation	Relevant findings
Research Collaboration Decisions	Bishop et al. 2014	Bishop, P. R., Huck, S. W., Ownley, B. H., Richards, J. K., and Skolits, G. J. (2014). Impacts of an interdisciplinary research center on participant publication and collaboration patterns: A case study of the National Institute for Mathematical and Biological Synthesis. *Research Evaluation*, 23(4), 327.	Case studies are important to understanding collaboration behaviors and effective decision-making in research collaborations
People in Research Groups			
Virtual Research Collaborations	Vanchieri et al. 2013	Vanchieri, T., Sebby, L., and Dooley, G. (2013). Toward a ubiquitous virtual collaboration environment: A fusion of traditional and leading-edge virtualization tools that empower distributed participants to explore, discover and exchange information without traditional boundaries or constraints. *Information Services and Use*, 33(3), 235–241.	Basic research initiatives and multidisciplinary university research collaborations can evolve through virtual centers.
Research Collaboration People Issues	Beaver 2001	Beaver, D. B. (2001). Reflections on scientific collaboration (and its study): Past, present, and future. *Scientometrics*, 52(3), 365–377.	Synergy, feedback, dissemination, recognition, and visibility between and among people in research collabora- tions can contribute to effective research collaborations.
Research Collaboration People Issues	Considine et al. 2014	Considine, M., Alexander, D., and Lewis, J. M. (2014). Policy design as craft: Teasing out policy design expertise using a semi-experimental approach. *Policy Sciences*, 47(3), 209–225.	Large and hierarchical organizations can be a challenge the role for less experienced or less powerful researchers.

Continued on next page

Collaboration topic	In-text citation	Full citation	Relevant findings
Research Collaboration People Issues	Chompalov et al. 2002	Chompalov, I., Genuth, J., and Shrum, W. (2002). The organization of scientific collabora- tions. *Research Policy*, 31(5), 749–767.	Collaborative projects are about people and not necessarily related to techno- logical practices.
Research Collaboration People Issues	Huang 2014	Haung, J. S. (2014). Building research collaboration networks: An interpersonal perspective for research capacity building. *Journal of Research Administration*, 45(2), 89–112.	Homophily in research interests among researchers is the most current ap- proach to research collaboration but not necessarily the best approach; heterophilous com- munications and maintaining degrees of heterophily between people in a collaboration can be challenging.
Research Collaboration People Issues	Youtie and Bozeman 2014	Youtie, J., and Bozeman, B. (2014). Social dynamics of research collaboration: norms, practices, and ethical issues in determining co-authorship rights. *Scientometrics*, 101(2), 953–962.	Most parties to research collaborations are satisfied with the results but when they are not, they are concerned about miscommunication about intellectual property or exploita- tion of contributors' work.
		Gender Dynamics	
Women in Research Collaborations	Corley and Gaughan 2005	Corley, E., and Gaughan, M. (2005). Scientists' par- ticipation in university research centers: What are the gender differenc- es? *Journal of Technology Transfer*, 30(4), 371–381.	Interdisciplinary centers may provide an institutional con- text where gender inequity in research collaborations can be achieved.

Collaboration topic	In-text citation	Full citation	Relevant findings
Women in Research Collaborations	Gaughan and Corley 2010	Gaughan, M., and Corley, E.A. (2010). Science faculty at US research universities: The impacts of university research center-affiliation and gender on industrial activities, *Technovation*, 30(3), 215–222.	Development of university research centers has advantages for both men and women. Male university research-center affiliates enjoy a slightly greater advantage than female center affiliates in their research collaboration outcomes.
Women in Research Collaborations	Abramo et al. 2013	Abramo, G., D'Angelo, C. A., and Murgia, G. (2013). Gender differences in research collaboration. *Journal of Informetrics*, 7811–7822.	An analysis of the scientific production of Italian academics shows that women researchers register a greater capacity to collaborate in all the forms analyzed, with the exception of international collaboration, where there is still a gap in comparison to male colleagues.
Women in Research Collaborations	Johnson and Bozeman 2012	Johnson J., and Bozeman B. (2012). Perspective: Adopting an asset bundles model to support and advance minority students' careers in academic medicine and the scientific pipeline. *Academic Medicine*, 87, 1488–1495.	The authors define "asset bundles" as the specific sets of abilities and resources individuals develop that help them succeed in science and research. This approach can be used to promote women in science.
Women in Research Collaborations	Pollak and Niemann 1998	Pollak, K. I., and Niemann, Y. F. (1998). Black and white tokens in academia: A difference of chronic versus acute distinctiveness. *Journal of Applied Social Psychology*, 28(11), 954–972.	Blacks and women differ in their research collaborations and related opportunities for academic performance.

Continued on next page

Collaboration topic	In-text citation	Full citation	Relevant findings
Women in Research Collaborations	Liao 2011	Liao, C.H. (2011) How to improve research quality? Examining the impacts of collaboration intensity and member diversity in collaboration networks. *Scientometrics*, 86(3), 747–761.	The authors demonstrate that when scholars are deeply embedded in a collaboration network, then race and gender are less important with respect to higher research quality.
Women in Research Collaborations	Etzkowitz et al. 2000	Etzkowitz, H., Kemelgor, C., Uzzi, B. (2000). *Athena Unbound: The Advancement of Women in Science and Technology*. Cambridge University Press, Cambridge, UK.	Two worlds of science still exist, one for male scientists and the other for female scientists.
Women in Research Collaborations	Long 2001	Long, J.S. (2001). *From Scarcity to Visibility: Gender Differences in the Careers of Doctoral Scientists and Engineers*. National Academy of Sciences, Washington, DC.	There have been advances in the entry of women into science and engineering. Yet a woman's ability to finance their education, as well as the salary differences between men and women academics, may inhibit academic careers of women and thus their ability to engage in ongoing research collaborations
Women in Research Collaborations	Rotbart et al. 2012	Rotbart, H. A., McMillen, D., Taussig, H., and Daniels, S. R. (2012). Assessing gender equity in a large academic department of pediatrics. *Academic Medicine*, 87(1), 98–104.	Women in the life sciences also are making strides in career advancement; but they are still lagging behind men in opportunities for academic contributions.

Collaboration topic	In-text citation	Full citation	Relevant findings
Women in Research Collaborations	Bozeman and Corley 2004	Bozeman, B., and Corley, E. (2004). Scientists' collaboration strategies: Implications for scientific and technical human capital. *Research Policy*, 33(4), 599–616.	Research collaboration effectiveness is determined by personal attributes, including gender.
Women in Research Collaborations	Bozeman and Gaughan 2011	Bozeman, B. and Gaughan, M. (2011). How do men and women differ in research collaborations? An analysis of the collaboration motives and strategies of academic researchers. *Research Policy*, 40(10), 1393–1402.	Research collaboration effectiveness is determined by personal attributes, especially gender.
Women in Research Collaborations	Bozeman et al. 2001	Bozeman, B., Dietz, J. S., and Gaughan, M. (2001). Scientific and technical human capital: An alternative model for research evaluation. International *Journal of Technology Management*, 22(7), 716–740.	Research collaboration effectiveness is determined by personal attributes, especially gender.
Women in Research Collaborations	Rijnsoever and Hessels 2011	Rijnsoever, F.J. and Hessels, L.K. (2011). Factors associated with disciplinary and interdisciplinary research collaboration. *Research Policy*, 40(3), 463–472.	Gender patterns in research and research collaborations are changing.
Women in Research Collaborations	Abramo et al. 2013	Abramo, G., D'Angelo, C. A., and Murgia, G. (2013). Gender differences in research collaboration. *Journal of Informetrics*, 7811–7822.	More research collaborative efforts are needed, and women may have as good or greater propensity for and capacity to collaborate in intramural and extramural research opportunities as compared to men.

Continued on next page

Collaboration topic	In-text citation	Full citation	Relevant findings
Women in Research Collaborations	Bozeman and Gaughan 2007	Bozeman, B., and Gaughan, M. (2007). Impacts of grants and contracts on academic researchers' interactions with industry. *Research Policy*, 36(5), 694–707.	An industrial involvement index provides insights on male-female differences in contributions to research and research collaboration outcomes.
Women in Research Collaborations	Lin and Bozeman 2006	Lin, M. W., and Bozeman, B. (2006). Researchers' industry experience and productivity in university–industry research centers: A "scientific and technical human capital" explanation. *Journal of Technology Transfer*, 31(2), 269–290.	An industrial involvement index provides insights on male-female differences in contributions to research and research collaboration outcomes.
		Institutional and Professional Contexts for Effective Research Collaboration	
Institutional and Professional Contexts	Lee 2000	Lee, Y. S. (2000). The sustainability of university-industry research collaboration: An empirical assessment. *Journal of Technology Transfer*, 25(2), 111–113.	Research collaboration is effective as a result of institutional or professional context
Institutional and Professional Contexts— Procedures and Processes	Lee and Bozeman 2005	Lee, S., and Bozeman, B. (2005). The impact of research collaboration on scientific productivity. *Social Studies of Science*, 35(5), 673–702.	University and industry research structures and cultures are different, as are collaborations that involve only academic scientists, in both basic and applied fields, and those that involve more complex university-industry partnerships.
Institutional and Professional Contexts— Procedures and Processes	Audretsch et al. 2002	Audretsch, D. B., Bozeman, B., Combs, K. L., Feldman, M., Link, A. N., Siegel, D. S., and Wessner, C. (2002). The economics of science and technology. *Journal of Technology Transfer*, 27(2), 155–203.	Universities with network ties to firms tend to have greater research and development productivity, mostly due to firm access to the human capital from faculty and students at the universities.

Collaboration topic	In-text citation	Full citation	Relevant findings
Institutional and Professional Contexts— Procedures and Processes	Hackett 2005	Hackett, E. J. (2005). Introduction to the special guest-edited issue on scientific collaboration. *Social Studies of Science*, 35(5), 667–672.	Political and managerial priorities vary by discipline and profession, and this can affect research productivity.
Institutional and Professional Contexts— Procedures and Processes	Glenna et al. 2011	Glenna, L. L., Welsh, R., Ervin, D., Lacy, W. B., and Biscotti, D. (2011). Commercial science, scientists' values, and university biotechnology research agendas. *Research Policy*, 40, 957–968.	Changes in commercialization policies have altered university research agendas and the values of scientists of different disciplines.
Institutional and Professional Contexts— Procedures and Processes	Katz 2000	Katz, J. S. (2000). Scale-independent indicators and research evaluation. *Science and Public Policy*, 27(1), 23–36.	The size of an organization matters for research collaboration effectiveness.
Institutional and Professional Contexts— Normative and Ethical Issues	Rennie 2001 Rennie and Flanagin 1994 Rennie et al. 2000	Rennie, D. (2001) Who did what? Authorship and contribution in 2001. *Muscle and Nerve*, 24(10), 1274–1277. Rennie, D. and Flanagin, A. (1994). Authorship! Authorship! Guests, ghosts, grafters, and the two-sided coin. *Journal of the American Medical Association*, 271(6), 469–471. Rennie, D., Flanagin, A, and Yank, Y. (2000). The contribution of authors. *Journal of the American Medical Association*, 284: 89–91.	Scientists in the biomedical areas of academia have been addressing institutional contexts of research collaborations for many decades.
Institutional and Professional Contexts— Normative and Ethical Issues	Wainwright et al. 2006	Wainwright, S. P., Williams, C., Michael, M., Farsides, B., Cribb, A. (2006). Ethical boundary-work in the embryonic stem cell laboratory. *Sociology of Health and Illness*, 28(6), 732–748.	Scientists in the biomedical areas of academia have been addressing institutional contexts of research collaborations for many decades.

Continued on next page

Collaboration topic	In-text citation	Full citation	Relevant findings
Institutional and Professional Contexts— Normative and Ethical Issues	Cohen et al. 2004	Cohen, M.B., Tarnow, E., and De Young, B.R. (2004). Coauthorship in pathology, a comparison with physics and a survey-generated and member-preferred authorship guideline. *MedGenMe*d, 63(1), 1–5.	Scientists in the biomedical areas of academia have been addressing institutional contexts of research collaborations for many decades.
Institutional and Professional Contexts— Normative and Ethical Issues	Shrum et al. 2001 Shrum et al. 2007	Shrum, W., Chompalov, I., and Genuth, J. (2001). Trust, conflict and performance in sci-entific collaborations. *Social Studies of Science*, 31(5), 681–730. Shrum, W., Genuth, J., and Chompalov, I. (2007). *Structures of Scientific Collaboration*. MIT Press, Cambridge, MA.	Research on the ethics and sociopolitical dynamics of scientif-ic collaboration for specific institutions or professions remains scarce.
Institutional and Professional Contexts— Normative and Ethical Issues	Devine et al. 2005	Devine, E.B., Beney, J., and Lisa A. Bero, L.A. (2005). Equity, account-ability, transparency: Implementation of the contributorship concept in a multi-site study. *American Journal of Pharmaceutical Education*, 69(4), 455–459.	Medical research has special hazards resulting from unethical behavior, in part because of its massive operation of clinical trials.
Institutional and Professional Contexts— Normative and Ethical Issues	Klingensmith and Anderson 2006	Klingensmith, M. E. and Anderson, K. A. (2006). Educational scholarship as a route to academic promotion: A depiction of surgical education scholars. *American Journal of Surgery*, 191(4), 533–537.	Medical research has special hazards resulting from unethical behavior, in part because of its massive operation of clinical trials.

Collaboration topic	In-text citation	Full citation	Relevant findings
Institutional and Professional Contexts—Normative and Ethical Issues	Harris and Lyon 2013	Harris, F., and Lyon, F. (2013). Transdisciplinary environmental research: Building trust across professional cultures. *Environmental Science and Policy*, 31, 109–119.	Researching complex topics are resulting in more transdisciplinary research collaboration, between research and industry and between different disciplinary fields and organizational contexts. Each group and discipline has its own norm for what constitutes an effective research collaboration.
		Contributorship Issues	
Contributorship	Merton 1968 Merton 1995	Merton, R. K. (1968) The Matthew effect in science. *Science*, 159(3810), 56–63. Merton, R. K. (1995) The Thomas theorem and the Matthew effect. *Social Forces*, 74(2), 379–422.	The "Matthew Effect" suggests that credit will inevitably be disproportionately given to more senior researchers, regardless of the particular nature or extent of their contribution and as compared to less well known or junior collaborators.
Contributorship	Levsky et al. 2007	Levsky, M. E. Rosin, A. Coon, T. P. Enslow, W. L. and Miller, M. A. (2007). A descriptive analysis of authorship within medical journals, 1995–2005. *Southern Medical Journal*, 100(4), 371–375.	There are growing concerns about authorship in medical journals, including honorary authorship, ghost authorship, duplicate and redundant publications, and, most important, authors' refusal to accept responsibility for their articles despite their readiness to accept credit for professional purposes.

Continued on next page

Collaboration topic	In-text citation	Full citation	Relevant findings
Contributorship	Heffner 1981	Heffner, A. G. (1981). Funded research, multiple authorship, and subauthorship collaboration in four disciplines. *Scientometrics*, 3(1), 5–12.	Researchers have considerable autonomy in their collaboration choices, and collaboration strategies are based in part of judgments about the conferring of coauthorship and status.
Contributorship	Rennie 2001 Rennie and Flanagin 1994 Rennie et al. 2000	Rennie, D. (2001) Who did what? Authorship and contribution in 2001. *Muscle and Nerve*, 24(10), 1274–1277. Rennie, D. and Flanagin, A. (1994). Authorship! Authorship! Guests, ghosts, grafters, and the two-sided coin. *Journal of the American Medical Association*, 271(6), 469–471. Rennie, D., Flanagin, A, and Yank, Y. (2000). The contribution of authors. *Journal of the American Medical Association*, 284: 89–91.	Rennie has been credited with developing the concept of contributorship and how it relates to authorship and research collaborations.
Contributorship	Marušić et al. 2011	Marušić, A., Bošnjak, L., and Jerončić, A. (2011). A systematic review of research on the meaning, ethics and practices of authorship across scholarly disciplines. *Plos ONE*, 6(9), 1–17.	Most studies of practices of authorship come from the biomedical and health sciences fields.
Contributorship	Youtie and Bozeman 2014	Youtie, J., and Bozeman, B. (2014). Social dynamics of research collaboration: Norms, practices, and ethical issues in determining co-authorship rights. *Scientometrics*, 101(2), 953–962.	Bad collaborations can range from miscommunication to legal disputes.

Collaboration topic	In-text citation	Full citation	Relevant findings
Contributorship	Moffatt 2011	Moffatt, B. (2011). Responsible authorship: Why researchers must forgo honorary authorship. *Accountability in Research: Policies and Quality Assurance*, 18(2), 76.	Honorary authorship is fairly widespread in the biomedical sciences but can range from being discouraged in other disciplines to being tagged as an unethical authorship practice.
Contributorship— Student Issues	Slaughter et al. 2002	Slaughter, S., Campbell, T., Folleman, M.H., and Morgan, E. (2002). The "traffic" in graduate students: Graduate students as tokens of exchange between academe and industry. *Science, Technology and Human Values*, 27(2), 282–313.	There is widespread concern about the possibilities for student exploitations in collaborations where students do not receive credit for their contribution to knowledge products
Contributorship— Student Issues	Baldini 2008	Baldini, N. (2008). Negative effects of university patenting: Myths and grounded evidence. *Scientometrics*, 75(20), 289–311.	This is one of the few systematic studies about the extent to which students are exploited in research collaborations.
Contributorship— Student Issues	Welsh et al. 2008	Welsh, R., Glenna, L., Lacy, W., and Biscotti, D. (2008). Close enough but not too far: Assessing the effects of university-industry research relationships and the rise of academic capitalism. *Research Policy*, 37(10), 1854–1864.	Collaborations rooted in industry-university partnerships often have beneficial effects for students including early publication, job offers, and mentoring.
Contributorship— Student Issues	Bozeman and Boardman 2014	Bozeman, B., and Boardman, C. (2014). Research Collaboration and Team Science: A State-of the-Art Review and Agenda, Springer Publishing, New York, NY.	Collaborations rooted in industry-university partnerships often have beneficial effects for students including early publication, job offers, and mentoring.

Continued on next page

Collaboration topic	In-text citation	Full citation	Relevant findings
		Better Research Collaborations	
Collaboration and Research Productivity	Pravdić and Oluić-Vuković 1986	Pravdić, N., and Oluić-Vuković, V. (1986). Dual approach to multiple authorship in the study of collaboration/scientific output relationship. *Scientometrics*, 10(5), 259–280.	Research collaboration enhances productivity of scientific knowledge.
Collaboration and Research Productivity	Lee and Bozeman 2005	Lee, S., and Bozeman, B. (2005). The impact of research collaboration on scientific productivity. *Social Studies of Science*, 35(5), 673.	Research collaboration enhances productivity of scientific knowledge.
Collaboration and Research Productivity	Wuchty et al. 2007	Wuchty, S., Jones, B. F., and Uzzi, B. (2007). The increasing dominance of teams in production of knowledge. *Science*, 316(5827), 1036.	Research collaboration enhances productivity of scientific knowledge.
Collaboration and Research Productivity	Huang 2014	Huang, J. S., (2014). Building research collaboration networks: An interpersonal perspective for research capacity building. *Journal of Research Administration*, 45(2), 89–112.	Building research collaborations can be challenging. Collaboration networks have a nonlinear effect on research productivity. Fostering heterophilous communications and maintaining degrees of heterophily can be contradicting and thus challenging. And building research collaboration networks proactively requires a shift of research management philosophy as well as invention of analytical tools for research management.

Collaboration topic	In-text citation	Full citation	Relevant findings
Collaboration and Research Productivity	Franklin et al. 2001	Franklin, S. J., Wright, M., and Lockett, A. (2001). Academic and surrogate entrepreneurs in university spin-out companies. *Journal of Technology Transfer*, 26(1), 127–141.	Research collaboration has beneficial effects with respect to scientific contributions.
Collaboration and Research Productivity	Shane 2004	Shane, S. A. (2004). *Academic Entrepreneurship: University Spinoffs and Wealth Creation.* Edward Elgar Publishing, Northampton, MA.	Research collaboration has beneficial effects with respect to scientific contributions.
Collaboration and Research Productivity	Dietz and Bozeman 2005	Dietz, J. S., and Bozeman, B. (2005). Academic careers, patents, and productivity: Industry experience as scientific and technical human capital. *Research Policy*, 34 (3), 349–367.	Research collaboration has beneficial effects with respect to scientific contributions.
Collaboration and Research Productivity	Link and Siegel 2005	Link, A. N., and Siegel, D. S. (2005). University-based technology initiatives: Quantitative and qualitative evidence. *Research Policy*, 34(3), 253–257.	Research collaboration has beneficial effects with respect to scientific contributions.
Collaboration and Research Productivity	Perkmann and Walsh 2009	Perkmann, M., and Walsh, K. (2009). The two faces of collaboration: Impacts of university-industry relations on public research. *Industrial and Corporate Change*, 18(6), 1033.	Research collaboration has beneficial effects with respect to scientific contributions.
Collaboration and Research Collaboration Effectiveness	Allen 1977	Allen, T. J. (1977). *Managing the Flow of Technology: Technology Transfer and the Dissemination of Technological Information with the R&D Organization.* MIT Press, Cambridge, MA.	Research collaboration effectiveness is essentially about relationships between researchers.

Continued on next page

Collaboration topic	In-text citation	Full citation	Relevant findings
Collaboration and Research Collaboration Effectiveness	Hagedoorn et al. 2000	Hagedoorn, J., Link, A. N., and Vonortas, N. S. (2000). Research partnership. *Research Policy*, 29(4-5), 567–586.	Research collaboration effectiveness is essentially about relationships between researchers and their partnerships.
Collaboration and Research Collaboration Effectiveness	Sonnewald 2007	Sonnenwald, D. H. (2007). Scientific collaboration. *Annual Review of Information Science and Technology*, 41(1), 643–681.	Research collaboration effectiveness is essentially about relationships between researchers and their partnerships.

Chapter One: Research Collaboration and Team Science

1. We define research collaboration as "the social processes whereby researchers come together jointly to deploy their human and social capital for the collective production of scientific and technical knowledge." By contrast, coauthorship simply refers to the formal listing of names on knowledge products (e.g., journal articles, conference papers, books). While research collaboration and coauthoring are closely related, collaborating does not require coauthorship (some collaborators are not credited as coauthors) and coauthorship does not require collaboration on research (some who are credited as coauthors do no research but receive credit as administrators, funders, out of charity or due to professional bullying, among other possibilities).

2. We have no evidence in our research showing that multiculturalism poses significant problems in STEM research collaborations when the multiple cultures in questions are domestic minorities and domestic majority team members. Nor could we find evidence in other systematic studies. This apparent lack of conflict could be a function of the low number of minority-majority research teams, the fact that minority academic researchers tend to have the same class origins of whites, or more optimistically, effective human relations.

3. For present purposes we do not distinguish between multidisciplinary and interdisciplinary, but we do discuss this distinction in later chapters. Suffice to say that interdisciplinary is a higher bar, at least in a team context, since it implies an integration not required by the multidisciplinary, which simply means the presence of multiple disciplines.

4. Here is an indicator of limited recognition of the relation of the science of team science and research collaboration studies: the most cited paper on science of team science (Stokols 2008) includes 111 references. Almost all journal references are to papers published in health and medical journals. None of the most cited articles on "research collaboration" is cited, nor are any of the journals in which most of the work on research collaboration appears (e.g., *Research Policy*, *Scientometrics*, *Journal of Technology Transfer*, *Social Studies of Science*). This limited relation between team science and research collaboration studies continues, but of late it shows some sign of diminishing (see Cooke and Hilton 2015).

5. Indeed, our book likewise takes the view that much of the general (nonscience, nonresearch) literature on collaborative teams is relevant, and we make extensive use of these findings.

6. The names here and throughout the book are fictitious. In all the quoted material here we provide a version extremely close to the original data, editing to delete names and identifying information and in some cases correcting grammar. However, we correct errors in grammar or usage only in those cases required to avoid ambiguity and confusion.

7. Yes, we would have preferred to know the specific article. However, in an earlier phase of this study we did ask respondents to respond to experiences related to one specific article that we had essentially picked at random (though none more than a few years back in time). We soon

found that most were reluctant to discuss any collaboration with "names named." Indeed, the response rate was too low to have yielded valid data.

8. Given that we oversampled females and certain departments, we reweighted results to reflect the population distribution as indicated in the NSF Survey of Doctorate Recipients 2006 (the most recently available survey).

Chapter Two: Routine and Not-So-Routine

1. The equivalence with Tolstoy's pithy epigram is not perfect, though. Our results indicate that the vast majority of research collaborations are happy, perhaps more often than is true of family relations.

2. We are grateful to an anonymous reviewer of this book, a like-minded individual obsessed with both metaphors and Russian literature. The reviewer suggests an "Ivan Illyich Principle" for research productivity. The principle proposes that research *productivity* is sometimes the enemy of research *quality*. The analogy could work: in Tolstoy's *Ivan Illyich* the protagonist is constantly in search of his rather pedestrian notion of the "good life" (e.g., possibly including the happily productive scientist, nose to the grindstone), realizing only when dying the wisdom of the "transcendental life," (e.g., possibly focused on a life of high-quality, high-impact science). We suppose the metaphor strains a bit when equating research impact with death. Still, those familiar with the publishing strategy of "the least publishable unit" will perhaps be sympathetic with this particular reading of *Ivan Illyich*.

3. Our cases: "The Delinquent Contributor" and "The Unplanned and Unwanted Honorary Authorship." We are trying to make a point here with these personal cases. However, these are the only cases in the book that are from our own experience. All the other cases and data included here are from our project data, not directly from our own research lives.

4. We note that some include social science researchers in the label STEM. We are not doing so here.

Chapter Four: Thinking Systematically about Research Effectiveness

1. Thus, the massive organization theory literature includes research and theory focused not only on a general concept of organization effectiveness (Rojas 2000) but also on dimensions of organizational effectiveness (e.g., Quinn and Rohrbaugh 1983; Sowa, Seldon. and Sandfort 2004; Hartnell, Ou, and Kinicki 2011).

2. We are confident in this claim because we had three project members, one faculty associate and two doctoral students code all the concepts mentioned in responses about good or effective collaboration.

Chapter Six: "Decision Making in Collaborative Research Teams"

1. Bibliometric studies are very useful, just for different things. What things? They shed light on such important factors as the growth in numbers of coauthors and coauthor teams (Wuchy et al. 2007), including "hyperauthorship" (Cronin 2001), the increasing globalization of coauthorship (Wagner and Leydesdorff 2005), and the rise of interdisciplinary research (Braun and Schubert 2003; Porter and Rafols 2009).

2. We could have asked researchers about an explicit randomly selected paper (indeed, we tried this in several pilots), but pilot results indicated that some researchers felt awkward about giving comments about an explicitly mentioned paper and its set of co-authors (that they would be "informing" on their friends and collaborators).

3. We asked respondents to rate these motivations on a 10-point scale, where 1 represents "not important at all" and 10 represents "extremely important."

4. National Science Foundation/National Center for Science and Engineering Statistics, NSF-NIH Survey of Graduate Students and Postdoctorates in Science and Engineering.

5. National Science Board 2016. *Science and Engineering Indicators 2016*. National Science Foundation, Arlington, VA, chapter 5.

6. Our data show that alphabetical order remains common only in economics.

7. These statements did not lend themselves to yes- or no-type answers. Thus the survey provided a ten-point scale to respondents. The scale ranged from 1 for strongly agree to 10 for strongly disagree.

8. The modal rating was a 6, and nearly one-third of respondents responded with a rating of 1, 2, or 3, suggesting strong agreement about the scientific quality of the article.

Chapter Seven: Enhancing the Effectiveness of Research Teams

1. https://www.teamsciencetoolkit.cancer.gov.
2. 42 U.S.C. 1862o-1, section 7009.

Appendix 1

1. We included the postdocs to reduce problems of recall about undesirable collaborations relating to less experienced researchers. However, few actually participated in the survey, so we do not generalize to that segment using survey data.

2. http://wayback.archive-it.org/5902/20160210151913/http://www.nsf.gov/statistics/doctorates/pdf/sed2006.pdf.

REFERENCES

Aad, G., Abbott, B., Abdallah, J., Abdinov, O., Aben, R., Abolins, M., Abulaiti, Y., . . . and Piqueras, D. A. (2015). Combined measurement of the Higgs Boson Mass in pp collisions at s = 7 and 8 TeV with the ATLAS and CMS Experiments. *Physical Review Letters*, 114(19), 191803.

Abramo, G., and D'Angelo, C. A. (2015). The relationship between the number of authors of a publication, its citations and the impact factor of the publishing journal: Evidence from Italy. *Journal of Informetrics*, 9(4), 746–761.

Abramo, G., D'Angelo, C. A., and Murgia, G. (2013). Gender differences in research collaboration. *Journal of Informetrics*, 7811–7822. doi:10.1016/j.joi.2013.07.002.

Abramo, G., D'Angelo, C. A., Di Costa, F., and Solazzi, M. (2011). The role of information asymmetry in the market for university–industry research collaboration. *Journal of Technology Transfer*, 36(1), 84–100.

ACS. American Chemical Society (2014). Ethical Guidelines to Publication of Chemical Research. http://pubs.acs.org/userimages/ContentEditor/1218054468605/ethics.pdf. Retrieved January 21, 2014.

Adler, N. E., and Stewart, J. (2010). Using team science to address health disparities: MacArthur network as case example. *Annals of the New York Academy of Sciences*, 1186(1), 252–260.

Aerts, K., and Schmidt, T. (2008). Two for the price of one? Additionality effects of R&D subsidies: A comparison between Flanders and Germany. *Research Policy*, 37(5), 806–822.

Ahlstrom, D., and Wang, L. C. (2009). Groupthink and France's defeat in the 1940 campaign. *Journal of Management History*, 15(2), 159–177.

Allen, T. J. (1977). *Managing the Flow of Technology: Technology Transfer and the Dissemination of Technological Information with the R&D Organization*. Cambridge, MA: MIT Press.

Ambos, T. C., Mäkelä, K., Birkinshaw, J., and D'Este, P. (2008). When does university research get commercialized? Creating ambidexterity in research institutions. *Journal of Management Studies*, 45(8), 1424–1447.

Annesley, T. M. (2011). Passing the paternité test. *Clinical Chemistry*, 57(9), 1239–1241.

Anonymous (2004). Authorship without authorization. *Nature Materials*, 3(11), 743.

Ashforth, B. (1994). Petty tyranny in organizations. *Human Relations*, 47(7), 755–778.

Aschhoff, B., and Grimpe, C. (2011). Localized norms and academics' industry involvement: The moderating role of age on professional imprinting. Unpublished paper downloaded February 3, 2012 from http://ftp.zew.de/pub/zew-docs/veranstaltungen/innovationpatenting2011/papers/Grimpe.pdf.

Audretsch, D. B., Bozeman, B., Combs, K. L., Feldman, M., Link, A. N., Siegel, D. S., and Wessner, C. (2002). The economics of science and technology. *Journal of Technology Transfer*, 27(2), 155–203.

Austin, A. E. (2002). Preparing the next generation of faculty: Graduate school as socialization to the academic career. *Journal of Higher Education*, 73(1), 94–122.

Austin, M. A., Hair, M. S., and Fullerton, S. M. (2012). Research guidelines in the era of large-scale collaborations: An analysis of genome-wide association study consortia. *American Journal of Epidemiology*, 175(9), 962–969.

Ayoko, O. B., Callan, V. J., and Härtel, C. E. (2003). Workplace conflict, bullying, and counter-productive behaviors. *International Journal of Organizational Analysis*, 11(4), 283–301.

Bailyn, L. (2003). Academic careers and gender equity: Lessons learned from MIT1. *Gender, Work and Organization*, 10(2), 137–153.

Baker, Beth. (2015). The Science of Team Science: An emerging field delves into the complexities of effective collaboration. *BioScience* 65(7), 639–644.

Baldini, N. (2008). Negative effects of university patenting: Myths and grounded evidence. *Scientometrics*, 75(2), 289–311.

Banks, M., and Milestone, K. (2011). Individualization, gender and cultural work. *Gender, Work and Organization*, 18(1), 73–89.

Bartoo, H., and Sias, P. M. (2004). When enough is too much: Communication apprehension and employee information experiences. *Communication Quarterly*, 52(1), 15–26.

Beaver, D. B. (2001). Reflections on scientific collaboration (and its study): Past, present, and future. *Scientometrics*, 52(3), 365–377.

———— (2004). Does collaborative research have greater epistemic authority? *Scientometrics*, 60(3), 399–408.

Becher, T. (1994). The significance of disciplinary differences. *Studies in Higher Education*, 19(2), 151–161.

Behrens, T. R., and Gray, D. O. (2001). Unintended consequences of cooperative research: Impact of industry sponsorship on climate for academic freedom and other graduate student outcome. *Research Policy*, 30(2), 179–199.

Bekkers, R., Duysters, G., and Verspagen, B. (2002). Intellectual property rights, strategic technology agreements and market structure: The case of GSM. *Research Policy*, 31(7), 1141–1161.

Bennett, L. M., and Gadlin, H. (2012). Collaboration and team science. *Journal of Investigative Medicine*, 60(5), 768–775.

Bercovitz, J., and Feldman, M. (2008). Academic entrepreneurs: Organizational change at the individual level. *Organization Science*, 19, 69–89.

Berger, J., Cohen, B. P., and Zelditch Jr, M. (1972). Status characteristics and social interaction. *American Sociological Review*, 37(3), 241–255.

Beyer, S., and Bowden, E. M. (1997). Gender differences in self-perceptions: Convergent evidence from three measures of accuracy and bias. *Personality and Social Psychology Bulletin*, 23(2), 157–172.

Biagioli, M. (1998). The instability of authorship: Credit and responsibility in contemporary biomedicine. *FASEB Journal*, 12(1), 3–16.

Bidault, F., and Hildebrand, T. (2014). The distribution of partnership returns: Evidence from co-authorships in economics journals. *Research Policy*, 43(6), 1002–1013.

Bikard, M., Murray, F., Gans J. S. (2015). Exploring trade-offs in the organization of scientific work: Collaboration and scientific reward. *Management Science*, 61, 1473–95.

Binz-Scharf, M. C., Kalish, Y., and Paik, L. (2015). Making science: New generations of collaborative knowledge production. *American Behavioral Science*, 59:531–47.

Bishop, P. R., Huck, S. W., Ownley, B. H., Richards, J. K., and Skolits, G. J. (2014). Impacts of an interdisciplinary research center on participant publication and collaboration patterns: A case study of the National Institute for Mathematical and Biological Synthesis. *Research Evaluation*, 23(4), 327.

Blaxter L., Hughes, C., and Tight, M. (2006). *How to Research*, 3rd ed. Maidenhead, Berkshire, UK: Open University Press.

Blumberg, B., Cooper, D. R., and Schindler, P. S. (2005). *Business Research Methods*, New York, NY: McGraw Hill.

Boardman, P. C., and Bozeman, B. (2006). Implementing a "bottom-up," multi-sector research collaboration: The case of the Texas air quality study. *Economics of Innovation and New Technology*, 15(1), 51–69.

——— (2007). Role strain in university research centers. *Journal of Higher Education*, 78(4), 430–463.

——— (2015). Academic faculty as intellectual property in university–industry research alliances. *Economics of Innovation and New Technology*, 24(5), 403–420.

Boardman, P. C., and Corley, E. A. (2008). University research centers and the composition of research collaborations. *Research Policy*, 37(5), 900–913.

Boardman, P. C., and Gray, D. (2010). The new science and engineering management: cooperative research centers as government policies, industry strategies, and organizations. *Journal of Technology Transfer*, 35(5), 445–459.

Boardman, P. C., and Ponomariov, B. L. (2007). Reward systems and NSF university research centers: The impact of tenure on university scientists' valuation of applied and commercially relevant research. *Journal of Higher Education*, 78(1), 51–70.

Bodas Freitas, I. M., Geuna, A., and Rossi, Federica (2013). Finding the right partners: Institutional and personal modes of governance of university–industry interactions. *Research Policy* 42 (1), 50–62.

Bordons, M., Aparicio, J., González-Albo, B., and Díaz-Faes, A. A. (2015). The relationship between the research performance of scientists and their position in co-authorship networks in three fields. *Journal of Informetrics*, 9(1), 135–144.

Börner, K., Contractor, N., Falk-Krzesinski, H. J., Fiore, S. M., Hall, K. L., Keyton, Spring, B., Stokols, D., Trochim, W. and Uzzi, B. (2010). A multi-level systems perspective for the science of team science. *Science Translational Medicine*, 2(49), 49cm24–49cm24.

Bozeman, B. (2000). Technology transfer and public policy: a review of research and theory. *Research Policy*, 29, 627–655.

Bozeman, B., and Boardman, C. (2004). The NSF Engineering Research Centers and the university–industry research revolution: A brief history featuring an interview with Erich Bloch. http://sites.nationalacademies.org/DBASSE/BBCSS/DBASSE. Retrieved May 2015.

———. (2014). Assessing research collaboration studies: A framework for analysis. In *Research Collaboration and Team Science*. Cham, Switzerland: Springer International Publishing.

Bozeman, B., and Corley, E. (2004). Scientists' collaboration strategies: Implications for scientific and technical human capital. *Research Policy*, 33(4), 599–616.

Bozeman, B., Dietz, J. S., and Gaughan, M. (2001). Scientific and technical human capital: An alternative model for research evaluation. *International Journal of Technology Management*, 22(7), 716–740.

Bozeman, B., Fay, D., and Slade, C. P. (2013). Research collaboration in universities and academic entrepreneurship: The-state-of-the-art. *Journal of Technology Transfer*, 38(1), 1–67.

Bozeman, B., Slade, C., and Hirsch, P. (2009). Understanding bureaucracy in health science ethics: Toward a better institutional review board. *American Journal of Public Health*, 99(9), 1549–1556.

Bozeman, B., and Gaughan, M. (2007). Impacts of grants and contracts on academic researchers' interactions with industry. *Research Policy*, 36(5), 694–707.

——— (2011). How do men and women differ in research collaborations? An analysis of the collaboration motives and strategies of academic researchers. *Research Policy*, 40(10), 1393–1402.

Bozeman, B., Gaughan, M., Youtie, J., Slade, C. P., and Rimes, H. (2015). Research collaboration experiences, good and bad: Dispatches from the front lines. *Science and Public Policy*, scv035. Online first, retrieved June 30, 2016.

Bozeman, B., Rimes, H., and Youtie, J. (2015). The evolving state-of-the-art in technology transfer research: Revisiting the contingent effectiveness model. *Research Policy*, 44(1), 34–49.

Bozeman, B., and Rogers, J. (2001). Strategic management of government-sponsored R&D portfolios. *Environment and Planning C: Government and Policy*, 19(3), 413–442.

———— (2002). A churn model of knowledge value: Internet researchers as a knowledge value collective. *Research Policy*, 31(4), 769–794.

Braun, T., and Schubert, A. (2003). A quantitative view on the coming of age of interdisciplinarity in the sciences 1980–1999. *Scientometrics*, 58(1), 183–189.

Brockner, J., and Wiesenfeld, B. M. (1996). An integrative framework for explaining reactions to decisions: Interactive effects of outcomes and procedures. *Psychological Bulletin*, 120(2), 189–208.

Brown, H. G., Poole, M. S., and Rodgers, T. L. (2004). Interpersonal traits, complementarity, and trust in virtual collaboration. *Journal of Management Information Systems*, 20(4), 115–138.

Bruneel, J., D'Este, P., and Salter, A. (2010). Investigating the factors that diminish the barriers to university–industry collaboration. *Research Policy*, 39(7), 858–868.

Bstieler, L., Hemmert, M., and Barczak, G. (2015). Trust formation in university–industry collaborations in the US biotechnology industry: Ip policies, shared governance, and champions. *Journal of Product Innovation Management*, 32(1), 111–121.

Buisseret, T. J., Cameron, H. M., and Georghiou, L. (1995). What difference does it make? Additionality in the public support of RD in large firms. *International Journal of Technology Management*, 10, 4(5), 587–600.

Burt, R. S. (2000). The network structure of social capital. *Research in Organizational Behavior*, 22, 345–423.

Caloghirou, Y., Tsakanikas, A., and Vonortas, N. S. (2001). University–industry cooperation in the context of the European framework programmes. *Journal of Technology Transfer*, 26(1), 153–161.

Carayannis, E. G., and Laget, P. (2004). Transatlantic innovation infrastructure networks: Public-private, EU–US R&D partnerships. *R&D Management*, 34(1), 17–31.

Carayol, N., and Matt, M., 2004. Does research organization influence academic production? Laboratory level evidence from a large European university. *Research Policy,* 33, 1081–1102.

Carr, P. L., Gunn, C. M., Kaplan, S. A., Raj, A., and Freund, K. M. (2015). Inadequate progress for women in academic medicine: Findings from the National Faculty Study. *Journal of Women's Health* (15409996), 24(3), 190–199. doi:10.1089/jwh.2014.4848.

Cassi, L., and Plunket, A. (2015). Research collaboration in co-inventor networks: Combining closure, bridging and proximities. *Regional Studies*, 49(6), 936–954.

Chang, D. B., and Dozier, K. (1995). Technology transfer and academic education with a focus on diversity. *Journal of Technology Transfer*, 20(3), 88–95.

Cheruvelil, K. S., Soranno, P. A., Weathers, K. C., Hanson, P. C., Goring, S. J., Filstrup, C. T., and Read, E. K. (2014). Creating and maintaining high-performing collaborative research teams: The importance of diversity and interpersonal skills. *Frontiers in Ecology and the Environment*, 12(1), 31–38.

Chompalov, I., Genuth, J., and Shrum, W. (2002). The organization of scientific collaborations. *Research Policy*, 31(5), 749–767.

Chompalov, I., and Shrum, W. (1999). Institutional collaboration in science: A typology of technological practice. *Science Technology and Human Values*, 24 (3), 338–372.

Clark, B. Y. (2011). Influences and conflicts of federal policies in academic–industrial scientific collaboration. *Journal of Technology Transfer*, 36(5), 514–545.

Clark, B. Y., and Llorens, J. J. (2012). Investments in scientific research: Examining the funding threshold effects on scientific collaboration and variation by academic discipline. *Policy Studies Journal*, 40:698–729.

Clark, M. S., and Mills, J. R. (2011). A theory of communal (and exchange) relationships. *Handbook of Theories of Social Psychology*. Los Angeles, CA: Sage Publications, 232–250.

Clarysse, B., Wright, M., and Mustar, P. (2009). Behavioural additionality of R&D subsidies: A learning perspective. *Research Policy*, 38(10), 1517–1533.

Claxton, L. D. (2005). Scientific authorship: Part 2. History, recurring issues, practices, and guidelines. *Mutation Research/Reviews in Mutation Research*, 589(1), 31–45.

Cohen, M.B., Tarnow, E., and De Young, B.R. (2004). Coauthorship in pathology, a comparison with physics and a survey-generated and member-preferred authorship guideline. *MedGenMed*, 63(1), 1–5.

Cohen, W. M., Nelson, R. R., and Walsh, J. P. (2002). Links and impacts: The influence of public research on industrial R&D. *Management Science*, 48(1), 1–23.

Colcleugh, D. (2013). *Everyone a Leader: A Guide to Leading High-Performance Organizations for Engineers and Scientists*. University of Toronto Press.

Collins, S., and Wakoh, H. (2000). Universities and technology transfer in Japan: Recent reforms in historical perspective. *Journal of Technology Transfer*, 25(2), 213–222.

Conrath, D. W. (1968). The role of the informal organization in decision making on research and development. *IEEE Transactions on Engineering Management*, 15(3), 109–119.

Considine, M., Alexander, D., and Lewis, J. M. (2014). Policy design as craft: Teasing out policy design expertise using a semi-experimental approach. *Policy Sciences*, 47(3), 209–225. doi:http://dx.doi.org.ezproxy.gru.edu/10.1007/s11077-013-9191-0.

Conte, M. L., Maat, S. L., and Omary, M. B. (2013). Increased co-first authorships in biomedical and clinical publications: A call for recognition. *FASEB Journal*, 27(10), 3902–3904.

Cooke, N. J., and Hilton, M. L. (eds.) (2015). *Enhancing the Effectiveness of Team Science*. Washington, DC: National Academies Press.

Cooper, M. H. (2009). Commercialization of the university and problem choice by academic biological scientists. *Science Technology Human Values*, 34(5), 629–653.

COPE (2002). Committee on Publication Ethics, *The COPE Report* 2002. http://www.publicationethics.org.uk/. Retrieved January 22, 2013.

Costas, R., and Bordons, M. (2011). Do age and professional rank influence the order of authorship in scientific publications? Some evidence from a micro-level perspective. *Scientometrics*, 88(1), 145–161.

Corey, G., Corey, M. S., Callanan, P., and Russell, J. M. (2014). *Group Techniques*. Stamford, CT: Cengage Learning.

Corley, E., and Gaughan, M. (2005). Scientists' participation in university research centers: What are the gender differences? *Journal of Technology Transfer*, 30(4), 371–381.

Corley, E. A., Boardman, P. C., and Bozeman, B. (2006). Design and the management of multi-institutional research collaborations: Theoretical implications from two case studies. *Research Policy*, 35(7), 975–993.

Council of Science Editors (2012). White paper on promoting integrity in scientific journal publications, 2012 update (approved by the CSE Board of Directors on March 30, 2012). http://www.councilscienceeditors.org/i4a/pages/index.cfm?pageid=3313. Retrieved January 22, 2014.

Cronin, B. (2001). Hyperauthorship: A postmodern perversion or evidence of a structural shift in scholarly communication practices? *Journal of the American Society for Information Science and Technology*, 52(7), 558–569.

Crow, M. M., and Bozeman, B. (1998). *Limited by design: R&D laboratories in the US national innovation system.* New York, NY: Columbia University Press.

Crowston, K., Specht, A., Hoover, C., Chudoba, K. M., and Watson-Manheim, M. B. (2015). Perceived discontinuities and continuities in transdisciplinary scientific working groups. *Science of the Total Environment*, 534, 159–172.

Cummings, J. N., and Kiesler, S. (2005). Collaborative research across disciplinary and organizational boundaries. *Social Studies of Science*, 35(5), 703.

Cummings, J. N., Kiesler, S., Zadeh, R. B., and Balakrishnan, A. D. (2013). Group Heterogeneity increases the risks of large group size: A longitudinal study of productivity in research groups. *Psychological Science*, 24(6), 880–890.

D'Este, P., and Perkmann, M. (2011). Why do academics engage with industry? The entrepreneurial university and individual motivations. *Journal of Technology Transfer*, 36(3), 316–339.

Davenport, S., Davies, J., and Grimes, C. (1998). Collaborative research programmes: Building trust from difference. *Technovation*, 19(1), 31–40.

Davis, L., Larsen, M. T., and Lotz, P. (2011). Scientists' perspectives concerning the effects of university patenting on the conduct of academic research in the life sciences. *Journal of Technology Transfer*, 36(1), 14–37.

Delbecq, A. L., Van de Ven, A. H., and Gustafson, D. H. (1975). *Group techniques for program planning: A guide to nominal group and Delphi processes.* Glenview, IL: Scott Foresman.

Dennis, A. R., Fuller, R. M., and Valacich, J. S. (2008). Media, tasks, and communication processes: A theory of media synchronicity. *MIS Quarterly*, 32(3), 575–600.

Devine, E. B., Beney, J., and Bero, L. A. (2005). Equity, accountability, transparency: Implementation of the contributorship concept in a multi-site study. *American Journal of Pharmaceutical Education*, 69(4), 455–459.

Dietz, J. S., and Bozeman, B. (2005). Academic careers, patents, and productivity: Industry experience as scientific and technical human capital. *Research Policy*, 34 (3), 349–367.

Diller, L. H. (2005). Fallout from the pharma scandals: The loss of doctors' credibility? *Hastings Center Report*, 35(3), 28–29.

Dirks, K. T. (1999). The effects of interpersonal trust on work group performance. *Journal of Applied Psychology*, 84(3), 445.

Dodgson, M. (1993). Learning, trust, and technological collaboration. *Human Relations*, 46(1), 77–95.

Dooley, L., and Kenny, B. (2015). Research collaboration and commercialization: The PhD candidate perspective. *Industry and Higher Education*, 29(2), 93–110.

Drenth, J. P. H. (1998). Multiple authorship: The contribution of senior authors. *Journal of the American Medical Association*, 280(3), 219–221.

Driskell, J. E., and Salas, E. (1991). Group decision making under stress. *Journal of Applied Psychology*, 76(3), 473.

Ductor, L. (2015). Does co-authorship lead to higher academic productivity? *Oxford Bulletin of Economics and Statistics*, 77(3), 385–407.

Duke, C. S., and Porter, J. H. (2013). The Ethics of Data Sharing and Reuse in Biology. *Bioscience*, 63(6), 483–489. doi:10.1525/bio.2013.63.6.10.

Duque, R. B., Ynalvez, M., Sooryamoorthy, R., Mbatia, P., Dzorgbo, D. B. S., and Shrum, W. (2005). Collaboration paradox. *Social Studies of Science*, 35(5), 755.

Egghe, L., Guns, R., and Rousseau, R. (2013). Measuring co-authors' contribution to an article's visibility. *Scientometrics*, 95(1), 55–67.

Eisenberg, R. L., Ngo, L. H., and Bankier, A. A. (2013). Honorary authorship in radiologic research articles: Do geographic factors influence the frequency? *Radiology*, 271(2), 472–478.

Elliott, C. (2014). Relationships between physicians and Pharma: Why physicians should not accept money from the pharmaceutical industry. *Neurology: Clinical Practice*, 4(2), 164–167.

Emerson, R. M. (1976). Social exchange theory. *Annual Review of Sociology*, 335–362.

Ennis, J. G. (1992). The social organization of sociological knowledge: Modeling the intersection of specialties. *American Sociological Review*, 57(2), 259–265.

Erickson, T. J., and Gratton, L. (2007). Eight ways to build collaborative teams. *Harvard Business Review*, 11, 1–11.

Eveline, J. (2004). *Ivory Basement Leadership: Power and Invisibility in the Changing University*. Seattle, WA: UWA Publishing.

Etzkowitz, H., Kemelgor, C., and Uzzi, B. (2000). *Athena Unbound: The Advancement of Women in Science and Technology*. Cambridge, UK: Cambridge University Press.

Falk-Krzesinski, H. J., Contractor, N., Fiore, S. M., Hall, K. L., Kane, C., Keyton, J., Spring, B., Stokols, D., Trochim, W. and Uzzi, B. (2011). Mapping a research agenda for the science of team science. *Research Evaluation*, 20(2), 145–158.

Faria, J. R., and Goel, R. K. (2010). Returns to networking in academia. *Netnomics*, 11(2), 103–117.

Feller, I. (1997). Federal and state government roles in science and technology. *Economic Development Quarterly*, 11(4), 283–295.

Feller, I., and Feldman, M. (2010). The commercialization of academic patents: Black boxes, pipelines, and Rubik's cubes. *Journal of Technology Transfer*, 35(6), 597–616.

Feller, I., and Roessner, D. (1995). What does industry expect from university partnerships? Congress wants to see bottom-line results from industry/government programs, but that's not what the participating companies are seeking. *Issues in Science and Technology*, 12(1) 80–84.

Fenwick, G. D., and Neal, D. J. (2001). Effect of gender composition on group performance. *Gender, Work and Organization*, 8(2), 205–225.

Fine, M. A., and Kurdel, L. A. (1993). Reflections on determining authorship credit and authorship order on faculty-student collaborations. *American Psychologist*, 48(11), 1141–1152.

Finholt, T. (2002). Collaboratories. *Annual Review of Information Science and Technology*, 36, 73–108.

Fink, A. (2010). *Conducting Research Literature Reviews: From the Internet toPaper*, 3rd ed. Thousand Oaks, CA: Sage Publishing.

Finkelstein, M. J., and Altbach, P. G. (eds.) (2014). *The Academic Profession: The Professoriate in Crisis*. New York, NY: Routledge.

Finkelstein, S., Whitehead, J., and Campbell, A. (2013). *Think Again: Why Good Leaders Make Bad Decisions and How to Keep it From Happening to You*. Cambridge, MA: Harvard Business Press.

Fiore, S. M. (2008). Interdisciplinarity as teamwork: How the science of teams can inform team science. *Small Group Research*, 39(3), 251–277.

Fiore, S. M., Carter, D. R., and Asencio, R. (2015). Conflict, trust, and cohesion: Examining affective and attitudinal factors in science teams. In *Team Cohesion: Advances in Psychological Theory, Methods and Practice* (271–301). Bingley, UK: Emerald Group Publishing Limited.

Fiske, S. T. (1993). Controlling other people: The impact of power on stereotyping. *American Psychologist*, 48(6), 621.

Flipse, S. M., van der Sanden, M. A., and Osseweijer, P. (2014). Setting up spaces for collaboration in industry between researchers from the natural and social sciences. *Science and Engineering Ethics*, 20(1), 7–22. doi:10.1007/s11948-013-9434-7.

Floyd, S. W., Schroeder, D. M., and Finn, D. M. (1994). Only if I'm first author: Conflict over credit in management scholarship. *Academy of Management Journal*, 37 (3), 734–745.

Fox, M. F. (2005). Gender, family characteristics, and publication productivity among scientists. *Social Studies of Science*, 35(1), 131–150.

Fox, M. F., and Faver, C. A. (1984). Independence and cooperation in research: The motivations and costs of collaboration. *Journal of Higher Education*, 55(3), 347–359.

Fox, M. F., and Ferri, V. C. (1992). Women, men, and their attributions for success in academe. *Social Psychology Quarterly*, 55(3), 257–271.

Fox, M. F., and Mohapta, S. (2007). Social-organizational characteristics of work and publication productivity among academic scientists in doctoral-granting departments. *Journal of Higher Education*, 78(5), 542–571.

Fox, M. F., and Stephan, P. E. (2001). Careers of young scientists: Preferences, prospects and realities by gender and field. *Social Studies of Science*, 31(1), 109–122.

Frandsen, T. F., and Nicolaisen, J. (2010). What is in a name? Credit assignment practices in different disciplines. *Journal of Informetrics*, 4(4), 608–617.

Frankel, M. S., and Bird, S. J. (2003). The role of scientific societies in promoting research integrity. *Science and Engineering Ethics*, 9(2), 139–140.

Franklin, S. J., Wright, M., and Lockett, A. (2001). Academic and surrogate entrepreneurs in university spin-out companies. *Journal of Technology Transfer*, 26(1), 127–141.

Freidson, E., and Rhea, B. (1963). Processes of control in a company of equals. *Social Problems*, 11, 119.

Funk, C. L., Barrett, K. A., and Macrina, F. L. (2007). Authorship and publication practices: Evaluation of the effect of responsible conduct of research instruction to postdoctoral trainees. *Accountability in Research*, 14(4), 269–305.

Galinsky, A. D., Magee, J. C., Inesi, M. E., and Gruenfeld, D. H. (2006). Power and perspectives not taken. *Psychological Science*, 17(12), 1068–1074.

Galison, P., Hevly, B. (eds.) (1992). *Big Science: The Growth of Large-Scale Research*. Palo Alto, CA: Stanford University Press.

Gardner, S. K. (2013). Paradigmatic differences, power, and status: A qualitative investigation of faculty in one interdisciplinary research collaboration on sustainability science. *Sustainability Science*, 8(2), 241–252.

Gardner, W. L., and Martinko, M. J. (1988). Impression management in organizations. *Journal of Management*, 14(2), 321–338.

Garg, K.C., and Padhi, P. (2001). A study of collaboration in laser science and technology. *Scientometrics*, 51(10), 415–427.

Garner, J., Porter, A. L., Borrego, M., Tran, E., and Teutonico, R. (2013). Facilitating social and natural science cross-disciplinarity: Assessing the human and social dynamics program. *Research Evaluation*, 22(2), 134–144.

Garrett-Jones, S., Turpin, T., and Diment, K. (2010). Managing competition between individual and organizational goals in cross-sector research and development centres. *Journal of Technology Transfer*, 35(5), 527–546.

Gaughan, M., and Bozeman, B. (2016). Using the prisms of gender and rank to interpret research collaboration power dynamics. *Social Studies of Science*, 0306312716652249.

Gaughan, M., and Corley, E. A. (2010). Science faculty at US research universities: The impacts of university research center-affiliation and gender on industrial activities, *Technovation*, 30(3), 215–222.

Gaughan, M., and Ponomariov, B. (2008) Faculty publication productivity, collaboration, and grants velocity: Using curricula vitae to compare center-affiliated and unaffiliated scientists. *Research Evaluation*, 17, 2, 103–110.

Gazni, A., and Didegah, F. (2011). Investigating different types of research collaboration and citation impact: A case study of Harvard University's publications. *Scientometrics*, 87(2), 251–265.

Geisler, E. (1986). The role of industrial advisory boards in technology transfer between universities and industry. *Journal of Technology Transfer*, 10(2), 33–42.

Gibbons, M., Limoges, C., Nowotny, H., Schwartzman, S., Scott, P., and Trow, M. (1994). *The New Production of Knowledge: The Dynamics of Science and Research in Contemporary Societies*. London, UK: Sage Publications.

Ginther, D. K., and Kahn, S. (2004). Women in economics: Moving up or falling off the academic career ladder? *Journal of Economic Perspectives*, 18, 193–214.

Glenna, L. L., Welsh, R., Ervin, D., Lacy, W. B., and Biscotti, D. (2011). Commercial science, scientists' values, and university biotechnology research agendas. *Research Policy*, 40(7), 957–968.

Godin, B. (1998). Writing performative history: The new New Atlantis? *Social Studies of Science*, 28(3), 465–483.

Goel, R. K., and Grimpe, C. (2011) Active versus passive academic networking: Evidence from micro-level data. *Journal of Technology Transfer*, 1–19.

Goldfarb, B., and Henrekson, M. (2003). Bottom-up versus top-down policies towards the commercialization of university intellectual property. *Research Policy*, 32(4), 639–658.

Gray, D. O. (2008). Making team science better: Applying improvement-oriented evaluation principles to evaluation of cooperative research centers. *New Directions for Evaluation*, 2008(118), 73–87.

——— (2011). Cross-sector research collaboration in the USA: A national innovation system perspective. *Science and Public Policy*, 38(2), 123–133.

Gray, D. O., and Steenhuis, H. J. (2003). Quantifying the benefits of participating in an industry university research center: An examination of research cost avoidance. *Scientometrics*, 58(2), 281–300.

Greene, M. (2007). The demise of the lone author. *Nature*, 450(7173), 1165.

Greenland, P., and Fontanarosa, P. B. (2012). Ending honorary authorship. *Science*, 337(6098), 1019.

Grimaldi, R., Kenney, M., Siegel, D. S., and Wright, M. (2011). 30 years after Bayh–Dole: Reassessing academic entrepreneurship. *Research Policy*, 40(8), 1045–1057.

Grimpe, C., and Fier, H. (2010). Informal university technology transfer: A comparison between the United States and Germany. *Journal of Technology Transfer*, 35 (6), 637–650.

Gross, C. (2016). Scientific misconduct. *Annual Review of Psychology*, 67, 693–711.

Grosse Kathoefer, D., and Leker, J. (2010), Knowledge transfer in academia: An exploratory study on the not-invented-here syndrome. *Journal of Technology Transfer*, 35(1),1–18.

Grossman, J. H., Reid, P. P., and Morgan, R. P. (2001). Contributions of academic research to industrial performance in five industry sectors. *Journal of Technology Transfer*, 26(1), 143–152.

Gruenfeld, D. H., Mannix, E. A., Williams, K. Y., and Neale, M. A. (1996). Group composition and decision making: How member familiarity and information distribution affect process and performance. *Organizational Behavior and Human Decision Processes*, 67(1), 1–15.

Guellec, D., and Van Pottelsberghe de la Potterie, B. (2004). From R&D to productivity growth: Do the institutional settings and the source of funds of R&D matter? *Oxford Bulletin of Economics and Statistics*, 66(3), 353–378.

Gulbrandsen, M., and Etzkowitz, H. (1999). Convergence between Europe and America: The transition from industrial to innovation policy. *Journal of Technology Transfer*, 24(2), 223–233.

Gulbrandsen, M., and Smedby, J. C. (2005). Industry funding and university professors' research performance. *Research Policy*, 34(6), 932–950.

Hackett, E. J. (2005). Introduction to the special guest-edited issue on scientific collaboration. *Social Studies of Science*, 35(5), 667–672.

Hackett, E. J., Conz, D., Parker, J., Bashford, J., and DeLay, S. (2004). Tokamaks and turbulence: Research ensembles, policy and technoscientific work. *Research Policy*, 33(5), 747–767.

Hackett, E. J., and Rhoten, D. R. (2009). The Snowbird Charrette: Integrative interdisciplinary collaboration in environmental research design. *Minerva*, 47(4), 407–440.

Haeussler, C., and Colyvas, J. A. (2011). Breaking the ivory tower: Academic entrepreneurship in the life sciences in UK and Germany. *Research Policy*, 40 (1), 41–54.

Haeussler, C., and Sauermann, H. (2013). Credit where credit is due? The impact of project contributions and social factors on authorship and inventorship. *Research Policy*, 42(3), 688–703.

——— (2014). The anatomy of teams: Division of labor and allocation of credit in collaborative knowledge production. Available at *SSRN 2434327*.

Hagedoorn, J., Link, A. N., and Vonortas, N. S. (2000). Research partnership. *Research Policy*, 29(4–5), 567–586.

Hagstrom, W. O. (1965) *The Scientific Community*. New York, NY: Basic Books.

Hall, B. H., Link, A. N., and Scott, J. T. (2001). Barriers inhibiting industry from partnering with universities: Evidence from the advanced technology program. *Journal of Technology Transfer*, 26(1), 87–98.

Hampton, S. E., and Parker, J. N. (2011). Collaboration and productivity in scientific synthesis. *BioScience*, 61(11), 900–910.

Hanel, P., and St-Pierre, M. (2006). Industry–University collaboration by Canadian manufacturing firms. *Journal of Technology Transfer*, 31(4), 485–499.

Hanson, S., and Pratt, G. (1995). *Gender, Work, and Space*. Psychology Press, London, UK.

Hara, N., Solomon, P., Kim, S. L., and Sonnenwald, D. H. (2003). An emerging view of scientific collaboration: Scientists' perspectives on collaboration and factors that impact collaboration. *Journal of the American Society for Information Science and Technology*, 54(10), 952–965.

Harris, F., and Lyon, F. (2013). Transdisciplinary environmental research: Building trust across professional cultures. *Environmental Science and Policy*, 31, 109–119. doi:10.1016/j.envsci.2013.02.006.

Harrison, T. R. (2004). What is success in ombuds processes? Evaluation of a university ombudsman. *Conflict Resolution Quarterly*, 21(3), 313–335.

Hartnell, C. A., Ou, A. Y., and Kinicki, A. (2011). Organizational culture and organizational effectiveness: A meta-analytic investigation of the competing values framework's theoretical suppositions. *Journal of Applied Psychology*, 96(4), 677.

Hearn, J. C., and Anderson, M. S. (2002). Conflict in academic departments: An analysis of disputes over faculty promotion and tenure. *Research in Higher Education*, 43(5), 503–529.

Heffner, A. G. (1981). Funded research, multiple authorship, and subauthorship collaboration in four disciplines. *Scientometrics*, 3(1), 5–12.

Hein, J., Zobrist, R., Konrad, C., and Schuepfer, G. (2012). Scientific fraud in 20 falsified anesthesia papers. *Der Anaesthesist*, 61(6), 543–549.

Heinze, T., and Bauer, G. (2007). Characterizing creative scientists in nano-S&T: Productivity, multidisciplinarity, and network brokerage in a longitudinal perspective. *Scientometrics*, 70(3), 811–830.

Herr, P. M. (1986). Consequences of priming: Judgment and behavior. *Journal of Personality and Social Psychology*, 51(6), 1106.

Hessels, L. K. (2013). Coordination in the science system: Theoretical framework and a case study of an intermediary organization. *Minerva: A Review of Science, Learning and Policy*, 51(3), 317–339. doi:10.1007/s11024-013-9230-1

Hessels, L. K., and Van Lente, H. (2008). Re-thinking new knowledge production: A literature review and a research agenda. *Research Policy*, 37(4), 740–760.

Hewstone, M. (1990). The 'ultimate attribution error'? A review of the literature on intergroup causal attribution. *European Journal of Social Psychology* 20.4: 311–335.

Hicks, D. M., and Katz, J. S. (1996). Where is science going? *Science, Technology and Human Values*, 21(4), 379.

Hillier, K. W. (1969). Educating scientists for industry. *Physics Bulletin*, 20(8), 335.

Hisrich, R. D., and Smilor, R. W. (1988). The university and business incubation: Technology transfer through entrepreneurial development. *Journal of Technology Transfer*, 13(1), 14–19.

Holland, D. (2013). *Integrating knowledge*. New York, NY: Routledge.

Holton, J. A. (2001). Building trust and collaboration in a virtual team. *Team Performance Management: An International Journal*, 7(3/4), 36–47.

Hou, H. Y., Kretschmer, H., and Liu, Z. Y. (2008). The structure of scientific collaboration networks in Scientometrics. *Scientometrics*, 75(2), 189–202.

Hoyt, C. L., and Burnette, J. L. (2013). Gender bias in leader evaluations merging implicit theories and role congruity perspectives. *Personality and Social Psychology Bulletin*, 39(10), 1306–1319.

Hsu, C. C., and Sandford, B. A. (2007). The Delphi technique: Making sense of consensus. *Practical Assessment, Research and Evaluation*, 12(10), 1–8.

Huang, D. W. (2015). Temporal evolution of multi-author papers in basic sciences from 1960 to 2010. *Scientometrics*, 105(3), 2137–2147.

Huang, J. S. (2014). Building research collaboration networks: An interpersonal perspective for research capacity building. *Journal of Research Administration*, 45(2), 89–112.

Huang, K. F., and Yu, C. M. J. (2011). The effect of competitive and non-competitive R&D collaboration on firm innovation. *Journal of Technology Transfer*, 36(4), 383–403.

Huang, M. H., and Lin, C. S. (2010*)*. International collaboration and counting inflation in the assessment of national research productivity. In *Proceedings of the American Society for Information Science and Technology*, 47(1), 1–4.

Hudson, J. M., Christensen, J., Kellogg, W. A., and Erickson, T. (April 2002). I'd be overwhelmed, but it's just one more thing to do: Availability and interruption in research management. In *Proceedings of the SIGCHI Conference on Human Factors in Computing Systems* . New York, NY: ACM Press, 97–104.

ICMJE International Committee of Medical Journal Editors (1997). Uniform requirements for manuscripts submitted to biomedical journals. *Journal of the American Medical Association*, 277(11), 927–934.

IJABE Guidelines for Authors. (2012). *International Journal of Agricultural and Biological Engineering*, 5(4), 96–103.

Inkpen, A. C., and Currall, S. C. (2004). The coevolution of trust, control, and learning in joint ventures. *Organization Science*, 15(5), 586–599.

Insel, T. R. (2010). Psychiatrists' relationships with pharmaceutical companies: Part of the problem or part of the solution? *Journal of the American Medical Association*, 303(12), 1192–1193.

Janis, I. L. (1982). *Groupthink: Psychological Studies of Policy Decisions and Fiascoes*. Vol. 349. Boston, MA: Houghton Mifflin.

Jankowski, J. E. (1999). Trends in academic research spending, alliances, and commercialization. *Journal of Technology Transfer*, 24(1), 55–68.

Jeffrey, P. (2003). Smoothing the waters: Observations on the process of cross-disciplinary research collaboration. *Social Studies of Science*, 33(4), 539–562.

Jeong, S., Choi, J. Y., and Kim, J. (2011). The determinants of research collaboration modes: Exploring the effects of research and researcher characteristics on co-authorship. *Scientometrics*, 89, 3, 967–983.

Jha, Y., and Welch, E. W. (2010). Relational mechanisms governing multifaceted collaborative behavior of academic scientists in six fields of science and engineering. *Research Policy*, 39(9), 1174–1184.

John, G., Cliff, N., Notarius, C., Markman, H., Bank, S., Yoppi, B., and Rubin, M. (1976). Behavior exchange theory and marital decision making. *Journal of Personality and Social Psychology*, 34(1), 14.

Johansson, M., Jacob, M., and Hellström, T. (2005). The strength of strong ties: University spin-offs and the significance of historical relations. *Journal of Technology Transfer*, 30(3), 271–286.

Johnson, J., and Bozeman, B. (2012). Perspective: Adopting an asset bundles model to support and advance minority students' careers in academic medicine and the scientific pipeline. *Academic Medicine*, 87, 1488–1495.

Johnson, W. H. A. (2009). Intermediates in triple helix collaboration: The roles of 4th pillar organisations in public to private technology transfer. *International Journal of Technology Transfer and Commercialisation*, 8(2), 142–158.

Jones, A. H. (2003). Can authorship policies help prevent scientific misconduct? What role for scientific societies? *Science and Engineering Ethics*, 9(2), 243–256.

Joshi, A. (2014). By whom and when is women's expertise recognized? The interactive effects of gender and education in science and engineering teams. *Administrative Science Quarterly*, published online before printing, March 2014. doi: 0001839214528331.

Kabanoff, B. (1991). Equity, equality, power, and conflict. *Academy of Management Review*, 16(2), 416–441.

Kant, L., Skogstad, A., Torsheim, T., and Einarsen, S. (2013). Beware the angry leader: Trait anger and trait anxiety as predictors of petty tyranny. *Leadership Quarterly*, 24(1), 106–124.

Katz, J. (1994). Geographical proximity and scientific collaboration. *Scientometrics*, 31(1), 31–43

Katz, J. S. (2000). Scale-independent indicators and research evaluation. *Science and Public Policy*, 27(1), 23–36.

Katz, J. S., and Hicks, D. (1997). How much is a collaboration worth? A calibrated bibliometric model. *Scientometrics*, 40(3), 541–554.

Katz, J. S., and Martin, B. R. (1997). What is research collaboration? *Research Policy*, 26(1), 1–18.

Kabo, F. W., Cotton-Nessler, N., Hwang, Y., Levenstein, M. C., Owen-Smith, J. 2014. Proximity effects on the dynamics and outcomes of scientific collaborations. *Research Policy* 43:1469–85.

Keeney, S., Hasson, F., and McKenna, H. (2006). Consulting the oracle: Ten lessons from using the Delphi technique in nursing research. *Journal of Advanced Nursing*, 53(2), 205–212.

Keltner, D., Gruenfeld, D. H., and Anderson, C. (2003). Power, approach, and inhibition. *Psychological Review*, 110(2), 265.

Kempers, R. D. (2002). Ethical issues in biomedical publications. *Fertility and Sterility*, 77(5), 883–888.

Kerr, W. R. (2013). *US High-Skilled Immigration, Innovation, and Entrepreneurship: Empirical Approaches and Evidence* (No. w19377). Cambridge, MA: National Bureau of Economic Research.

Kilduff, G. J., Willer, R., and Anderson, C. (2016). Hierarchy and its discontents: Status disagreement leads to withdrawal of contribution and lower group performance. *Organization Science*, 27(2), 373–390.

Klingensmith, M. E., and Anderson, K. A. (2006). Educational scholarship as a route to academic promotion: A depiction of surgical education scholars. *American Journal of Surgery*, 191(4), 533–537.

Klotz, A. C., Hmieleski, K. M., Bradley, B. H., and Busenitz, L. W. (2014). New venture teams a review of the literature and roadmap for future research. *Journal of Management*, 40(1), 226–255.

König, B., Diehl, K., Tscherning, K., and Helming, K. (2013). A framework for structuring interdisciplinary research management. *Research Policy*, 42(1), 261–272.

Kovacs, J. (2013). Honorary authorship epidemic in scholarly publications? How the current use of citation-based evaluative metrics make (pseudo) honorary authors from honest contributors of every multi-author article. *Journal of Medical Ethics*, 39(8), 509–512.

Kraemer, S. (2006). *Science and technology policy in the United States: Open systems in action.* New Brunswick, NJ: Rutgers University Press.

Kuhn, T. S. (1996). *TheSstructure of Scientific Revolutions.* Chicago, IL: University of Chicago Press.

Kyvik, S. (2013). The academic researcher role: Enhancing expectations and improved performance. *Higher Education*, 65(4), 525–538.

Kyvik, S., and Teigen, M. (1996). Child care, research collaboration, and gender differences in scientific productivity. *Science, Technology and Human Values*, 21(1), 54–71.

Lagnado, M. (2003a). Increasing the trust in scientific authorship. *British Journal of Psychiatry*, 183(1), 3–4.

———— (2003b). Professional writing assistance: Effects on biomedical publishing. *Learned Publishing*, 16(1), 21–27.

Lakhani, J., Benzies, K., and Hayden, K. A. (2012). Attributes of interdisciplinary research teams: A comprehensive review of the literature. *Clinical and Investigative Medicine*, 35(5), 260–265.

Landry, R., and Amara, N. (1998). The impact of transaction costs on the institutional structuration of collaborative academic research. *Research Policy* 27, 901–913.

Landry, R. J., Amara, N., and Ouimet, M. (2007). Determinants of knowledge transfer: Evidence from Canadian university researchers in natural sciences and engineering. *Journal of Technology Transfer*, 32(6), 561–592.

Laudel, G. (2001). Collaboration, creativity and rewards: Why and how scientists collaborate. *International Journal of Technology Management*, 22, 762–81.

Le, T., and Gardner, S. K. (2010). Understanding the doctoral experience of Asian international students in the science, technology, engineering, and mathematics (STEM) fields: An exploration of one institutional context. *Journal of College Student Development*, 51(3), 252–264.

Leahey, E. (2006). Gender differences in productivity: Research specialization as a missing link. *Gender and Society*, 20, 754–780.

———— (2007). Convergence and confidentiality? Limits to the implementation of mixed methodology. *Social Science Research*, 36, 149–158.

———— (2016). From solo investigator to team scientist: Trends in the practice and study of research collaboration. *Annual Review of Sociology*, 42, 81–100.

Leahey, E., Beckman, C., and Stanko, T. (2015). Prominent but less productive: The impact of interdisciplinarity on scientists' research. arXiv:1510.06802 [cs.DL].

Leahey, E., and Moody, J. (2014). Sociological innovation through subfield integration. *Social Currents*, 1, 28–56

Leahey, E., and Reikowsky, R. 2008. Research specialization and collaboration patterns in sociology. *Social Studies of Science*, 38, 425–440.

Lederer, S., and Davis, A. B. (1995). Subjected to science: Human experimentation in America before the Second World War. *History: Reviews of New Books*, 24(1), 13–13.

Ledford, H. (2015). Team science. *Nature*, 525(7569), 308–311.

Lee, S., and Bozeman, B. (2005). The impact of research collaboration on scientific productivity. *Social Studies of Science,* 35(5) 673–702.

Lee, Y. N., Walsh, J. P., and Wang, J. (2015). Creativity in scientific teams: Unpacking novelty and impact. *Research Policy,* 44(3), 684–697.

Lee, Y. S. (2000). The sustainability of university–industry research collaboration: An empirical assessment. *Journal of Technology Transfer*, 25(2), 111–133.

Lester, J. (2008). Performing gender in the workplace: Gender socialization, power, and identity among women faculty members. *Community College Review*, 35(4), 277–305.

Levi, D. (2016). *Group dynamics for teams.* Los Angeles, CA: Sage Publications.

Levine, J. M., and Moreland, R. L. (1994). Group socialization: Theory and research. *European Review of Social Psychology*, 5(1), 305–336.

Levsky, M. E., Rosin, A., Coon, T. P., Enslow, W. L., and Miller, M. A. (2007). A descriptive analysis of authorship within medical journals, 1995–2005. *Southern Medical Journal,* 100, 371–375.

Leyden, D. P., and Link, A. N. (1999). Federal laboratories as research partners. *International Journal of Industrial Organization*, 17(4), 575–592.

Lcvy, R., Roux, P., and Wolff, S. (2009). An analysis of science–industry collaborative patterns in a large European university. *Journal of Technology Transfer*, 34(1), 1–23.

Lewis, J. M., Ross, S., and Holden, T. (2012). The how and why of academic collaboration: Disciplinary differences and policy implications. *Higher Education*, 64(5), 693–708.

Li, E. Y., Liao, C. H., and Yen, H. R. (2013). Co-authorship networks and research impact: A social capital perspective. *Research Policy*, 42(9), 1515–1530.

Liao, C. H. (2011) How to improve research quality? Examining the impacts of collaboration intensity and member diversity in collaboration networks. *Scientometrics,* 86(3), 747–761.

Liao, C. H., and Yen, H. R. (2012). Quantifying the degree of research collaboration: A comparative study of collaborative measures. *Journal of Informetrics*, 6(1), 27–33.

Lightman, B. (ed.) (2008). *Victorian Science in Context.* Chicago, IL: University of Chicago Press.

Lin, M. W., and Bozeman, B. (2006). Researchers' industry experience and productivity in university–industry research centers: A "scientific and technical human capital" explanation. *Journal of Technology Transfer*, 31(2), 269–290.

Link, A. N., and Siegel, D. S. (2005). University-based technology initiatives: Quantitative and qualitative evidence. *Research Policy*, 34(3), 253–257.

Link, A. N., Siegel, D. S., and Bozeman, B. (2007). An empirical analysis of the propensity of academics to engage in informal university technology transfer, industrial and corporate change. *Research Policy*, 16(4), 641–655.

Linstone, H. A., and Turoff, M. (eds.) (1975). *The Delphi Method: Techniques and Applications.* Reading, MA: Addison-Wesley.

Liu, H., Chang, B., and Chen, K. (2012). Collaboration patterns of Taiwanese scientific publications in various research areas. *Scientometrics*, 29(1), 145–165.

Liyanage, S., and Mitchell, H. (1994). Strategic management of interactions at the academic-industry interface. *Technovation*, 14(10), 641–655.

Long, J. S. (2001). *From Scarcity to Visibility: Gender Differences in the Careers of Doctoral Scientists and Engineers.* Washington, DC: National Academy of Sciences.

Long, J. S., and Fox. M. F. (1995). Scientific careers: Universalism and Particularism. *Annual Review of Sociology*, 21, 45–71.

Lööf, H., and Broström, A. (2008). Does knowledge diffusion between university and industry increase innovativeness? *Journal of Technology Transfer*, 33(1), 73–90.

Lotrecchiano, G. R. (2013). A dynamical approach toward understanding mechanisms of team science: Change, kinship, tension, and heritage in a transdisciplinary team. *Clinical and Translational Science*, 6(4), 267–278.

Lounsbury, J. W., Foster, N., Patel, H., Carmody, P., Gibson, L. W., and Stairs, D. R. (2012). An investigation of the personality traits of scientists versus nonscientists and their relationship with career satisfaction. *R&D Management*, 42(1), 47–59.

Lowell, B. L. (2010). A long view of America's immigration policy and the supply of foreign-born STEM workers in the United States. *American Behavioral Scientist*, 53(7), 1029–1044.

Lozano, G. A. 2013. The elephant in the room: Multi-authorship and the assessment of individual researchers. *Current Science* 105(4): 443–445

Luukkonen, T. (2000). Additionality of EU framework programmes 1. *Research Policy*, 29(6), 711–724.

Lynch, J., Strasser, J. E., Lindsell, C. J., and Tsevat, J. (2013). Factors that affect integrity of authorship of scientific meeting abstracts. *AJOB Primary Research*, 4(2), 15–22. doi:10.1080/21507716.2012.757259.

Maass, A., and Clark, R. D. (1984). Hidden impact of minorities: Fifteen years of minority influence research. *Psychological Bulletin*, 95(3), 428.

Madden, J. F. (1998). Authorship credit in student-faculty publications [authorship credit—almanac 44 (21) 1998 1–3]. http://www.upenn.edu/almanac/v44/n21/authorship.html. Accessed January 20, 2014.

Madlock, P. E., and Martin, M. M. (2012). Communication and work alienation: To speak or not to speak. *Human Communication*, 14(4), 369–382.

Mandler, G. (1980). Recognizing: The judgment of previous occurrence. *Psychological Review*, 87(3), 252.

Mangematin, V., O'Reilly, P., and Cunningham, J. (2014). PIs as boundary spanners, science and market shapers. *Journal of Technology Transfer*, 39(1), 1–10.

Manolio, T. A., Rodriguez, L. L., Brooks, L., Abecasis, G., Ballinger, D., Daly, M., . . . and Collins, F. S. (2007). New models of collaboration in genome-wide association studies: The Genetic Association Information Network. *Nature Genetics*, 39(9), 1045–1051.

Mansfield, E. (1995). Academic research underlying industrial innovations: Sources, characteristics, and financing. *Review of Economics and Statistics*, 77(1), 55–65.

Marks, G., and Miller, N. (1987). Ten years of research on the false-consensus effect: An empirical and theoretical review. *Psychological Bulletin*, 102(1), 72.

Martinelli, A., Meyer, M., and Tunzelmann, N. (2008). Becoming an entrepreneurial university? A case study of knowledge exchange relationships and faculty attitudes in a medium-sized, research-oriented university. *Journal of Technology Transfer*, 33(3), 259–283.

Marušić, A., Bošnjak, L., and Jerončić, A. (2011). A systematic review of research on the meaning, ethics and practices of authorship across scholarly disciplines. *Plos ONE*, 6(9), 1–17. doi:10.1371/journal.pone.0023477.

Marusic, M., Bozikov, J., Katavic, V., Hren, D., Kljaković-Gaspić, M., and Marusić, A. (2004). Authorship in a small medical journal: A study of contributorship statements by corresponding authors. *Science and Engineering Ethics*, 10(3), 493–502.

Mâsse, L. C., Moser, R. P., Stokols, D., Taylor, B. K., Marcus, S. E., Morgan, G. D., Hall, K.L., Croyle, R.T., and Trochim, W. M. (2008). Measuring collaboration and transdisciplinary integration in team science. *American Journal of Preventive Medicine*, 35(2), S151–S160.

Mastroianni, A. C., and Kahn, J. P. (2002). Risk and responsibility: Ethics, Grimes v Kennedy Krieger, and public health research involving children. *American Journal of Public Health*, 92(7), 1073–1076.

Matt, M., Robin, S., and Wolff, S. (2011). The influence of public programs on inter-firm R&D collaboration strategies: Project-level evidence from EU FP5 and FP6. *Journal of Technology Transfer*, 37(6), 885–916. doi: 10.1007/s10961-011-9232-9.

Mattsson, P., Laget, P., Nilsson, A., and Sundberg, C. (2008). Intra-EU vs. extra-EU scientific co-publication patterns in EU. *Scientometrics*, 75(3), 555–574.

Mayrose, I., and Freilich, S. (2015). The interplay between scientific overlap and cooperation and the resulting gain in co-authorship interactions. *Plos ONE*, 10(9), 1–10. doi:10.1371/journal.pone.0137856.

McCrary, S., Anderson, C., Jakovljevic, J., Khan, T., McCullough, L., Wray, N., and Brody, B. (2000). A national survey of policies on disclosure of conflicts of interest in biomedical research. *New England Journal of Medicine*, 343(22), 1621–1626.

McGrath, J. E., and Hollingshead, A. B. (1994). *Groups Interacting with Technology: Ideas, Evidence, Issues, and an Agenda*. Los Angeles, CA: Sage Publications.

McKelvey, B., and Sekaran, U. (1977). Toward a career-based theory of job involvement: A study of scientists and engineers. *Administrative Science Quarterly*, 281–305.

McKelvey, M., Zaring, O., and Ljungberg, D. (2015). Creating innovative opportunities through research collaboration: An evolutionary framework and empirical illustration in engineering. *Technovation*, 39, 26–36.

McKenna, H. P. (1994). The Delphi technique: A worthwhile research approach for nursing? *Journal of Advanced Nursing*, 19(6), 1221–1225.

McMillan, S. S., Kelly, F., Sav, A., Kendall, E., King, M. A., Whitty, J. A., and Wheeler, A. J. (2014). Using the Nominal Group Technique: How to analyse across multiple groups. *Health Services and Outcomes Research Methodology*, 14(3), 92–108.

Melin, G. (2000). Pragmatism and self-organization: Research collaboration on the individual level. *Research Policy*, 29(1), 31–40.

Melin, G., and Persson, O. (1996). Studying research collaboration using co-authorships. *Scientometrics*, 36(3), 363–377.

Mendoza, P. (2007). Academic capitalism and doctoral student socialization: A case study. *Journal of Higher Education*, 78(1), 71–96.

Merton, R. K. (1968). The Matthew effect in science. *Science*, 159(3810), 56–63.

——— (1995). The Thomas theorem and the Matthew effect. *Social Forces*, 74(2), 379–422.

Meyer, M. (2006). Academic inventiveness and entrepreneurship: On the importance of start-up companies in commercializing academic patents. *Journal of Technology Transfer*, 31(4), 501–510.

Millum, J., and Menikoff, J. (2010). Streamlining ethical review. *Annals of Internal Medicine*, 153(10), 655–657.

Mitchell, G. E., O'Leary, R., and Gerard, C. (2015). Collaboration and performance: Perspectives from public managers and NGO leaders. *Public Performance and Management Review*, 38(4), 684–716. doi:10.1080/15309576.2015.1031015.

Moffatt, B. (2011). Responsible authorship: Why researchers must forgo honorary authorship. *Accountability in Research: Policies and Quality Assurance*, 18(2), 76. doi:10.1080/0898962 1.2011.557297.

Monge, P. R., Cozzens, M. D., and Contractor, N. S. (1992). Communication and motivational predictors of the dynamics of organizational innovation. *Organization Science*, 3(2), 250–274.

Moody, J. (2004). The structure of a social science collaboration network: Disciplinary cohesion from 1963 to 1999. *American Sociological Review*, 69(2), 213–238.

Morgan, R. P., Kruytbosch, C., and Kannankutty, N. (2001). Patenting and invention activity of US scientists and engineers in the academic sector: Comparisons with industry. *Journal of Technology Transfer*, 26(1), 173–183.

Morgan, R. P., and Strickland, D. E. (2001). U.S. university research contributions to industry. *Science and Public Policy*, 28(2), 113–122.

Mosvick, R. K. (1971). Human relations training for scientists, technicians, and engineers: A review of relevant experimental evaluations of human relations training. *Personnel Psychology*, 24(2), 275–292.

Mowatt, G., Shirran, L., Grimshaw, J. M., Rennie, D., Flanagin, A., Yank, V., Graeme, M., Gretsche, P. C., and Bero, L. A. (2002). Prevalence of honorary and ghost authorship in *Cochrane Reviews*. *Journal of the American Medical Association*, 287(21), 2769–2771.

Mowery, D. C., and Sampat, B. N. (2001). Patenting and licensing university inventions: Lessons from the history of the research corporation. *Industrial and Corporate Change*, 10(2), 317–355.

Muchnik, L., Aral, S., and Taylor, S. J. (2013). Social influence bias: A randomized experiment. *Science*, 341(6146), 647–651.

Mullen, P. D., and Ramirez, G. (2006). The promise and pitfalls of systematic reviews. *Annual Review of Public Health*, 27, 81–102.

Müller, R. R. (2014). Postdoctoral life scientists and supervision work in the contemporary university: A case study of changes in the cultural norms of science. *Minerva: A Review of Science, Learning and Policy*, 52(3), 329–349. doi:10.1007/s11024-014-9257-y.

National Academies Committee on Science, Engineering, and Public Policy, Committee on Facilitating Interdisciplinary Research (2005). *Facilitating Interdisciplinary Research*. Washington, DC: National Academies Press.

National Academy of Engineering (2003). *The Impact of Academic Research on Industrial Performance*. Washington, DC: National Academies Press.

National Institutes of Health. Guidelines for the conduct of research in the intramural research program at NIH. http://sourcebook.od.nih.gov/ethic-conduct/Conduct%20Research%20 6-11-07.pdf. Accessed July 2, 2016.

National Science Foundation (2015a). Immigrants play increasing role in U.S. science and engineering workforce. News release 15-120. https://www.nsf.gov/news/news_summ .jsp?cntn_id=136430. Accessed July 2, 2016.

——— (2015b). Science and engineering degrees, by race/ethnicity of recipients: 2002–12, https://www.nsf.gov/statistics/2015/nsf15321/#chp1&chp2. Accessed July 3, 2016.

Nedeva, M., Georghiou, L., and Halfpenny, P. (1999). Benefactors or beneficiary: The role of industry in the support of university research equipment. *Journal of Technology Transfer*, 24(2), 139–147.

Nemeth, C. J. (1986). Differential contributions of majority and minority influence. *Psychological Review*, 93(1), 23.

Nicholas, W. (2006). One last question: Who did the work? *New York Times*, 2.

Nilsson, A. S., Rickne, A., and Bengtsson, L. (2010). Transfer of academic research: Uncovering the grey zone. *Journal of Technology Transfer*, 35(6), 617–636.

Niosi, J. (2006). Introduction to the symposium: Universities as a source of commercial technology. *Journal of Technology Transfer*, 31(4), 399–402.

Öberg, G. (2009). Facilitating interdisciplinary work: Using quality assessment to create common ground. *Higher Education*, 57(4), 405–415.

O'Connor, G. C., Rice, M. P., Peters, L., and Veryzer, R. W. (2003). Managing interdisciplinary, longitudinal research teams: Extending grounded theory-building methodologies. *Organization Science*, 14(4), 353–373.

Olson, G., and Olson, J. S. (2000). "Distance matters," *Human Computer Interaction*. 15(2), 139–178.

Osborne, J. W., and Holland, A. (2009). What is authorship, and what should it be? A survey of prominent guidelines for determining authorship in scientific publications. *Practical Assessment, Research and Evaluation*, 14(15), 1–19.

Parker, C. (2015). Practicing conflict resolution and cultural responsiveness within interdisciplinary contexts: A study of community service practitioners. *Conflict Resolution Quarterly*, 32(3), 325–357.

Parker, J. N., and Hackett, E. J. (2012). Hot spots and hot moments in scientific collaborations and social movements. *American Sociological Review*, 77(1), 21–44.

Paul, S., Samarah, I. M., Seetharaman, P., and Mykytyn, P. P., Jr. (2004). An empirical investigation of collaborative conflict management style in group support system-based global virtual teams. *Journal of Management Information Systems*, 21(3), 185–222.

Pavitt, Charles (1999).Theorizing about the group communication-leadership relationship. *The Handbook of Group Communication Theory and Research*, 313–334.

Pennington, D. D., Simpson, G. L., McConnell, M. S., Fair, J. M., and Baker, R. J. (2013). Transdisciplinary research, transformative learning, and transformative science. *Bioscience*, 63(7), 564–573.

Perkmann, M., Tartari, V., McKelvey, M., Autio, E., Broström, A., D'Este, P., Fini, R., Geuna, A., Grimaldi, R., Hughes, A. and Krabel, S. (2013). Academic engagement and commercialisation: A review of the literature on university–industry relations. *Research Policy*, 42(2), 423–442.

Perkmann, M., and Walsh, K. (2009). The two faces of collaboration: Impacts of university–industry relations on public research. *Industrial and Corporate Change*, 18(6), 1033.

Peri, G., Shih, K., and Sparber, C. (2015). STEM workers, H-1B visas, and productivity in US cities. *Journal of Labor Economics*, 33(S1, pt. 2), S225–S255.

Peterson, R. S. (1997). A directive leadership style in group decision making can be both virtue and vice: Evidence from elite and experimental groups. *Journal of Personality and Social Psychology*, 72(5), 1107.

Pichini, S., Pulido, M., and García-Algar, O. (2005). Authorship in manuscripts submitted to biomedical journals: An author's position and its value. *Science and Engineering Ethics*, 11(2), 173–175.

Pieterse, A. N., Van Knippenberg, D., and Van Dierendonck, D. (2013). Cultural diversity and team performance: The role of team member goal orientation. *Academy of Management Journal*, 56(3), 782–804.

Plosila, W. H. (2004). State science- and technology-based economic development policy: History, trends and developments, and future directions. *Economic Development Quarterly*, 18(2), 113–126.

Pohl, C., Wuelser, G., Bebi, P., Bugmann, H., Buttler, A., Elkin, C., and . . . Huber, R. (2015). How to successfully publish interdisciplinary research: Learning from an Ecology and Society Special Feature. *Ecology and Society*, 20(2), 276–291. doi:10.5751/ES-07448-200223.

Pollak, K. I., and Niemann, Y. F. (1998). Black and white tokens in academia: A difference of chronic versus acute distinctiveness. *Journal of Applied Social Psychology*, 28(11), 954–972.

Ponds, R. (2009). The limits to internationalization of scientific research collaboration. *Journal of Technology Transfer*, 34(1), 76–94.

Ponomariov, B., and Boardman, P. C. (2008). The effect of informal industry contacts on the time university scientists allocate to collaborative research with industry. *Journal of Technology Transfer*, 33(3), 301–313.

——— (2010). Influencing scientists' collaboration and productivity patterns through new institutions: University research centers and scientific and technical human capital. *Research Policy*, 39, 613–624.

Ponomariov, B. L. (2008). Effects of university characteristics on scientists' interactions with the private sector: An exploratory assessment. *Journal of Technology Transfer*, 33, 485–503.

Porter, A., and Rafols, I. (2009). Is science becoming more interdisciplinary? Measuring and mapping six research fields over time. *Scientometrics*, 81(3), 719–745.

Powell, D. A., Jacob, C. J., and Chapman, B. J. (2012). Using blogs and new media in academic practice: Potential roles in research, teaching, learning, and extension. *Innovative Higher Education*, 37(4), 271–282.

Poyago-Theotoky, J., Beath, J., and Siegel, D. S. (2002). Universities and fundamental research: Reflections on the growth of university–industry partnerships. *Oxford Review of Economic Policy*, 18(1), 10–21.

Pravdić, N., and Oluić-Vuković, V. (1986). Dual approach to multiple authorship in the study of collaboration/scientific output relationship. *Scientometrics*, 10(5), 259–280.

Probert, B. (2005). "I just couldn't fit it in": Gender and unequal outcomes in academic careers. *Gender, Work and Organization*, 12(1), 50–72.

Qin, J., Lancaster, F. W., and Allen, B. (1997). Types and levels of collaboration in interdisciplinary research in the sciences. *Journal of the American Society for Information Science*, 48(10), 893–916.

Quinn, R. E., and Rohrbaugh, J. (1983). A spatial model of effectiveness criteria: Towards a competing values approach to organizational analysis. *Management Science*, 29(3), 363–377.

Raelin, J. A. (1989). An anatomy of autonomy: Managing professionals. *The Academy of Management Executive*, 3(3), 216–228.

Rai, A. K. (1999). Regulating scientific research: Intellectual property rights and the norms of science. *Northwestern University Law Review*, 94, 77.

Rappert, B., Webster, A., and Charles, D. (1999). Making sense of diversity and reluctance: Academic–industrial relations and intellectual property. *Research Policy*, 28(8), 873–890.

Regalado, A. (1995). Multiauthor papers on the rise. *Science* 268(5207), 25.

Renault, C. S. (2006). Academic capitalism and university incentives for faculty entrepreneurship. *Journal of Technology Transfer*, 31(2), 227–239.

Rennie, D. (2001) Who did what? Authorship and contribution in 2001. *Muscle and Nerve*, 24(10), 1274–1277.

Rennie, D., and Flanagin, A. (1994). Authorship! Authorship! Guests, ghosts, grafters, and the two-sided coin. *Journal of the American Medical Association*, 271(6), 469–471.

Rennie, D., Flanagin, A., and Yank, V. (2000). The contributions of authors. *Journal of the American Medical Association*, 284, 89-91.

Rennie, D., Yank, V., and Emanuel, L. (1997). When authorship fails: A proposal to make contributors accountable. *Journal of the American Medical Association,* 278(7), 579–585.

Rhoades, G., and Slaughter, S. (1997). Academic capitalism, managed professionals, and supply-side higher education. *Academic Labor*, 51, 9–38.

Rijnsoever, F. J., and Hessels, L. K. (2011). Factors associated with disciplinary and interdisciplinary research collaboration. *Research Policy*, 40(3), 463–472.

Riker, W. H. (1986). *The Art of Political Manipulation*. Vol. 587. New Haven, CT: Yale University Press,.

Roessner, D., Ailes, C. P., Feller, I., and Parker, L. (1998). How industry benefits from NSF's Engineering Research Centers. *Research Technology Management*, 41(5), 40.

Rogers, J. D. (2012). Research centers as agents of change in the contemporary academic landscape: Their role and impact in HBCU, EPSCoR, and Majority universities. *Research Evaluation*, 21(1), 15–32.

Rohrbaugh, J. (1981). Improving the quality of group judgment: Social judgment analysis and the nominal group technique. *Organizational Behavior and Human Performance*, 28(2), 272–288.

Rojas, R. R. (2000). A review of models for measuring organizational effectiveness among for-profit and nonprofit organizations. *Nonprofit Management and Leadership*, 11(1), 97–104.

Rossini, F. A., and Porter, A. L. (1981). Interdisciplinary research: Performance and policy issues. *Journal of the Society of Research Administrators*, 13 (2), 8–24.

Rotbart, H. A., McMillen, D., Taussig, H., and Daniels, S. R. (2012). Assessing gender equity in a large academic department of pediatrics. *Academic Medicine*, 87(1), 98–104.

Rothaermel, F. T., Agung, S. D., and Jiang, L. (2007). University entrepreneurship: A taxonomy of the literature. *Industrial and Corporate Change*, 16(4), 691–791.

Ryan, J. C. (2014). The work motivation of research scientists and its effect on research performance. *R&D Management*, 44(4), 355–369.

Rylance, R. (2015). Global funders to focus on interdisciplinarity. *Nature*. 525, 313–315. doi: 10.1038/525313a. pmid:26381969.

Sagie, A. (1996). Effects of leader's communication style and participative goal setting on performance and attitudes. *Human Performance*, 9(1), 51–64.

Salazar, M. R., Lant, T. K., Fiore, S. M., and Salas, E. (2012). Facilitating innovation in diverse science teams through integrative capacity. *Small Group Research*, 43(5), 527–558.

Sampat, B. N. (2006). Patenting and US academic research in the 20th century: The world before and after Bayh-Dole. *Research Policy*, 35(6), 772–789.

Saragossi, S., and van Pottelsberghe de la Potterie, B. (2003). What patent data reveal about universities: The case of Belgium. *Journal of Technology Transfer*, 28(1), 47–51.

Schachter, H. L. (1989). *Frederick Taylor and the Public Administration Community: A Reevaluation*. Albany, NY: SUNY Press.

Schartinger, D., Schibany, A., and Gassler, H. (2001). Interactive relations between universities and firms: Empirical evidence for Austria. *Journal of Technology Transfer*, 26(3), 255–268.

Schmidt, P. (2013). New test to measure faculty collegiality produces some dissension itself. *Chronicle of Higher Education*, June 10, 2013.

Schut, M., van Paassen, A., Leeuwis, C., and Klerkx, L. (2014). Towards dynamic research configurations: A framework for reflection on the contribution of research to policy and innovation processes. *Science and Public Policy*, 41(2), 207.

Selznick, P. (1943). An approach to a theory of bureaucracy. *American Sociological Review*, 8(1), 47–54.

Senécal, C., Julien, E., & Guay, F. (2003). Role conflict and academic procrastination: A self-determination perspective. *European Journal of Social Psychology*, 33(1), 135–145.

Shane, S. A. (2004). *Academic Entrepreneurship: University Spinoffs and Wealth Creation*. Northampton, MA: Edward Elgar Publishing.

Sherwood, A. L., and Covin, J. G. (2008). Knowledge acquisition in university–industry alliances: An empirical investigation from a learning theory perspective. *Journal of Product Innovation Management*, 25(2), 162–179.

Shrum, W., Chompalov, I., and Genuth, J. (2001). Trust, conflict and performance in scientific collaborations. *Social Studies of Science*, 31(5), 681–730.

Shrum, W., Genuth, J., and Chompalov, I. (2007). *Structures of Scientific Collaboration*. Cambridge, MA: MIT Press.

Siegel, D. S., Waldman, D. A., Atwater, L. E., and Link, A. N. (2003). Commercial knowledge transfers from universities to firms: Improving the effectiveness of university–industry collaboration. *Journal of High Technology Management Research*, 14(1), 111–133.

——— (2004). Toward a model of the effective transfer of scientific knowledge from academicians to practitioners: Qualitative evidence from the commercialization of university technologies. *Journal of Engineering and Technology Management*, 21, 115–142.

Siegel, D. S., Waldman, D. and Link, A. (2003). Assessing the impact of organizational practices on the relative productivity of university technology transfer offices: An exploratory study. *Research Policy,* 32, 27–48.

Slaughter, S., Campbell, T., Folleman, M. H., and Morgan, E. (2002). The "traffic" in graduate students: Graduate students as tokens of exchange between academe and industry. *Science, Technology and Human Values,* 27(2), 282–313.

Slaughter, S., and Leslie, L. L. (1997). *Academic Capitalism: Politics, Policies, and the Entrepreneurial University*. Baltimore, MD: Johns Hopkins University Press.

Slaughter, S., and Rhoades, G. (2004). *Academic capitalism and the new economy: Markets, state, and higher education*. Baltimore, MD: Johns Hopkins University Press.

Smith, S. (1985). Groupthink and the hostage rescue mission. *British Journal of Political Science*, 15(01), 117–123.

Sonnenfeld, J. A. (1985). Shedding light on the Hawthorne studies. *Journal of Organizational Behavior*, 6(2), 111–130.

Sonnenwald, D. H. (2007). Scientific collaboration. *Annual Review of Information Science and Technology*, 41(1), 643–681.

Sowa, J. E., Selden, S. C., and Sandfort, J. R. (2004). No longer unmeasurable? A multidimensional integrated model of nonprofit organizational effectiveness. *Nonprofit and Voluntary Sector Quarterly*, 33(4), 711–728.

Spector, P. E. (1986). Perceived control by employees: A meta-analysis of studies concerning autonomy and participation at work. *Human Relations*, 39(11), 1005–1016.

Spratlen, L. P. (1995). Interpersonal conflict which includes mistreatment in a university workplace. *Violence and Victims*, 10(4), 285–297.

Steen, R. G., Casadevall, A., and Fang, F. C. (2013). Why has the number of scientific retractions increased?. *PLoS One*, 8(7), e68397.

Steneck, N. H. (2006). Fostering integrity in research: Definitions, current knowledge, and future directions. *Science and Engineering Ethics*, 12, 53–74.

Stetten, D. (ed.). (1984). *NIH: An Account of Research in Its Laboratories and Clinics*. Orlando, FL: Academic Press.

Stokes, T. D., and Hartley, J. A. (1989). Coauthorship, social structure and influence within specialties. *Social Studies of Science*, 19(1), 101–125.

Stokols, D., Hall, K. L., Taylor, B. K., and Moser, R. P. (2008). The science of team science: Overview of the field and introduction to the supplement. *American Journal of Preventive Medicine*, 35(2), S77–S89.

Stuart, T. E., and Ding, W. W. (2006). When do scientists become entrepreneurs? The social structural antecedents of commercial activity in the academic life sciences. *American Journal of Sociology*, 112 (1), 97–144.

Suh, K. S. (1999). Impact of communication medium on task performance and satisfaction: An examination of media-richness theory. *Information and Management*, 35(5), 295–312.

Tartari, V., and Breschi, S. (2012). Set them free: Scientists' evaluations of the benefits and costs of university–industry research collaboration. *Industrial and Corporate Change*, 21(5), 1117–1147.

Tartari, V., and Salter, A. (2015). The engagement gap: Exploring gender differences in University–Industry collaboration activities. *Research Policy*, 44(6), 1176–1191.

Terpstra, D. E., and Honoree, A. L. (2004). Job satisfaction and pay satisfaction levels of university faculty by discipline type and by geographic region. *Education*, 124(3), 528.

Thursby, J. G., Jensen, R., and Thursby, M. C. (2001). Objectives, characteristics and outcomes of university licensing: A survey of major US universities. *Journal of Technology Transfer*, 26(1), 59–72.

Thursby, M., Thursby, J., and Gupta-Mukherjee, S. (2007). Are there real effects of licensing on academic research? A life cycle view. *Journal of Economic Behavior and Organization*, 63(4), 577–598.

Thurow, A. P., Abdalla, C. W., Younglove-Webb, J., and Gray, B. (1999). The dynamics of multidisciplinary research teams in academia. *Review of Higher Education*, 22(4), 425–440.

Toivanen, H., and Ponomariov, B. (2011) African regional innovation systems: Bibliometric analysis of research collaboration patterns. *Scientometrics*, 88(2), 471–493.

Tollefsen, D. P. (2006). Group deliberation, social cohesion, and scientific teamwork: Is there room for dissent? *Episteme*, 3(1–2), 37–51.

Treise, D., Baralt, C., Birnbrauer, K., Krieger, J., and Neil, J. (2016). Establishing the need for health communication research: Best practices model for building transdisciplinary collaborations. *Journal of Applied Communication Research*, 44(2), 194–198.

Turpin, T., Garrett-Jones, S., and Woolley, R. (2011) Cross-sector research collaboration in Australia: The Cooperative Research Centres Program at the Crossroads. *Science and Public Policy*, 38(2), 87–97.

Ubfal, D., and Maffioli, A. (2011). The impact of funding on research collaboration: Evidence from a developing country. *Research Policy*, 40(9), 1269–1279.

Ulnicane, I. (2015). Why do international research collaborations last? Virtuous circle of feedback loops, continuity and renewal. *Science and Public Policy*, 42(4), 433. doi:10.1093/scipol/scu060.

Van Noorden R. (2015). Interdisciplinary research by the numbers. *Nature*, 525, 306–307. doi:10.1038/525306a. pmid:26381967.

Van Rijnsoever, F. J., and Hessels, L. K. (2011). Factors associated with disciplinary and interdisciplinary research collaboration. *Research Policy*, 40(3), 463–472.

Vangen, S., and Huxham, C. (2003). Nurturing collaborative relations: Building trust in interorganizational collaboration. *Journal of Applied Behavioral Science*, 39(1), 5–31.

Vanchieri, T., Sebby, L., and Dooley, G. (2013). Toward a ubiquitous virtual collaboration environment: A fusion of traditional and leading-edge virtualization tools that empower distributed participants to explore, discover and exchange information without traditional boundaries or constraints. *Information Services and Use*, 33(3), 235–241. doi:10.3233/ISU-130716.

Vasileiadou, E. (2012). Research teams as complex systems: Implications for knowledge management. *Knowledge Management of Research Practice*, 10(2), 118–127.

Vaughan, D. (1997). *The Challenger Launch Decision: Risky Technology, Culture, and Deviance at NASA*. Chicago, IL: University of Chicago Press.

Venkatramen, V. (2010). Conventions of scientific authorship. *ScienceCareers*, April 16.

Vinkler, P. (1993). Research contribution, authorship and team cooperativeness. *Scientometrics*, 26(1), 213–230.

Wagner, C., Brahmakulam, I., Jackson, B., Wong, A., and Yoda, T., 2001. *Science and Technology Collaboration: Building Capacity in Developing Countries?* MR-1357.0-WB. RAND, Santa Monica, CA.

Wagner, C. S. (2005). Six case studies of international collaboration in science. *Scientometrics*, 62(1), 3–26.

Wagner, C. S., and Leydesdorff, L. (2005). Network structure, self-organization, and the growth of international collaboration in science. *Research Policy*, 34(10), 1608–1618.

Wainwright, S. P., Williams, C., Michael, M., Farsides, B., and Cribb, A. (2006). Ethical boundary-work in the embryonic stem cell laboratory. *Sociology of Health and Illness*, 28(6), 732–748.

Walsh, J. P., Cohen, W. M., and Cho, C. (2007). Where excludability matters: Material versus intellectual property in academic biomedical research. *Research Policy*, 36(8), 1184–1203.

Walsh, J. P., and Lee, Y. N. (2015). The bureaucratization of science. *Research Policy*, 44(8), 1584–1600.

Waltman, L. (2012). An empirical analysis of the use of alphabetical authorship in scientific publishing. *Journal of Informetrics*, 6(4), 700–711.

Wang, J., and Shapira, P. (2011). Funding acknowledgement analysis: An enhanced tool to investigate research sponsorship impacts: The case of nanotechnology. *Scientometrics*, 87(3), 563–586.

Wasem, R. E. (May 2012). Immigration of foreign nationals with Science, Technology, Engineering, and Mathematics (STEM) degrees. Congressional Research Service, Washington, DC: Library of Congress.

Weinberg, S. L., and Scott, M. A. (2013). The impact of uncapping of mandatory retirement on postsecondary institutions. *Educational Researcher*, 42(6), 338–348.

Welsh, R., Glenna, L., Lacy, W., and Biscotti, D. (2008). Close enough but not too far: Assessing the effects of university–industry research relationships and the rise of academic capitalism. *Research Policy*, 37(10), 1854–1864.

Weltzin, J. F., Belote, R. T., Williams, L. T., Keller, J. K., and Engel, E. C. (2006). Authorship in ecology: Attribution, accountability, and responsibility. *Frontiers in Ecology and the Environment*, 4(8), 435–441.

White, R. F. (2007). Institutional review board mission creep: The common rule, social science, and the nanny state. *Independent Review*, 11(4), 547–564.

Williams, R. L., Zyzanski, S. J., Flocke, S. A., Kelly, R. B., and Acheson, L. S. (1998). Critical success factors for promotion and tenure in family medicine departments. *Academic Medicine*, 73(3), 333–335.

Wislar, J. S., Flanagin, A., Fontanarosa, P. B., and DeAngelis, C. D. (2011). Honorary and ghost authorship in high impact biomedical journals: A cross sectional survey. *British Medical Journal*, 343, d6128.

Wrege, C. D., and Perroni, A. G. (1974). Taylor's pig-tale: A historical analysis of Frederick W. Taylor's pig-iron experiments. *Academy of Management Journal*, 17(1), 6–27.

Wuchty, S., Jones, B. F., and Uzzi, B. (2007). The increasing dominance of teams in production of knowledge. *Science*, 316(5827), 1036.

Ynalvez, M., and Shrum, W. (2011) Professional networks, scientific collaboration, and publication productivity in resource-constrained research institutions in a developing country. *Research Policy*, 40(2), 204–216.

Younglove-Webb, J., Thurow, A. P., Abdalla, C. W., and Gray, B. (1999). The dynamics of multidisciplinary research teams in academia. *Review of Higher Education*, 22(4), 425–440.

Youtie, J., and Bozeman, B. (2014). Social dynamics of research collaboration: Norms, practices, and ethical issues in determining co-authorship rights. *Scientometrics*, 101(2), 953–962.

——— (2016). Dueling co-authors: How collaborators create and sometimes solve contributorship conflicts. *Minerva*, 5(4), 375–397.

Zega, M., D'Agostino, F., Bowles, K. H., De Marinis, M. G., Rocco, G., Vellone, E., and Alvaro, R. (2014). Development and validation of a computerized assessment form to support nursing diagnosis. *International Journal of Nursing Knowledge*, 25(1), 22–29.

Zerhouni, E. (2003). The NIH roadmap. *Science*, 302(5642), 63–72.

Zitt, Bassecoulard, E., and Okubo, Y. (2000). Shadows of the past in international cooperation: Collaboration profiles of the top five producers of science. *Scientometrics*, 47, 627–657.

Zucker, D. (2012). Developing your career in an age of team science. *Journal of Investigative Medicine*, 60(5), 779–784.

INDEX

academic capitalism and entrepreneurship, 3, 6–7, 66–67; suppression of results and, 141–42. *See also* commercial application of research; industry-university research collaborations

African Americans, 9

Aggregate Model of Research Collaboration Effectiveness, 28–29, 70, 72–79; collaboration management and, 29, 73–74, 78–79, 105; diagram of, 73; interviews of researchers and, 86, 105; strategies for addressing problems and, 130–42, 155

alphabetical author order, 5, 27, 111–12, 113

anonymous Web posts, 130, 159, 162

Asians and Asian Americans, 9

Assumptive Collaboration Management, 15, 142, 143, 144, 147–48, 155, 157

authorship problems. *See* coauthorship; ghost authors; honorary authorship

bad collaborations. *See* Nightmare Collaborations; Routinely Bad Collaborations

Bayh-Dole Act, 6

bibliometric studies: of discipline-based coauthorship norms, 111–12; of Italian women academics, 60; lacking evidence of effectiveness, 106; in research collaboration studies, 14, 58; uses of, 196n1. *See also* citation counts

biomedical research: bad interdisciplinary experiences with medical researchers, 89–90; clinical trials in, 62, 85; commercial activities of medical school faculty, 97; contributorship issues in, 63–64, 137; ethical issues of collaboration in, 62, 84–85, 187–89; ghost authors in, 5, 6, 64, 85, 118; honorary authorship in, 5, 64, 65, 116, 118; increased number of women in, 7; nightmare authorship-crediting problem in, 45–46; in propositional literature table, 187–88; team science studies focused on, 13, 14. *See also* medical

journals; National Institutes of Health (NIH); pharmaceutical industry

career stage: collaboration effectiveness and, 96–99; power dynamics related to, 76–77, 98–99, 130, 134; productivity of collaborating researcher and, 107, 130; strategies for problems associated with, 130, 131, 134–36. *See also* students; tenure

childbearing, 78, 102, 103, 104. *See also* maternity leave; pregnancy

citation counts: different approaches to calculation of, 4, 61. *See also* bibliometric studies

clinical trials, 62, 85

coauthorship: collaboration choices and, 64; concepts of collaboration not limited to, 51, 52, 53, 195n1; decision-making for, 111–15, 124; explicit discussions of collaborators about, 114–15, 124, 132, 156; order of listing authors and, 4–5, 27, 111–14; of patrons, 52; research collaboration defined as, 51, 53, 175–76; without actual research collaboration, 195n1. *See also* contributorship; crediting issues; ghost authors; honorary authorship; most recent coauthored publication; number of coauthors

coauthorship problems: career stage and, 130, 131, 134–36; disciplinary differences and, 133, 140–41; with interdisciplinary or multi-institutional research, 88; in most recent coauthored publication, 120–21; need for studies outside of biomedical sciences, 118; with person who made no contribution, 131; in respondent's whole career, 121–24; unwanted coauthorship, 41–42, 132, 137; of women negotiating for authorship positions, 133, 139

collaboration. *See* research collaboration

A NOTE ON THE TYPE

This book has been composed in Adobe Text and Gotham. Adobe Text, designed by Robert Slimbach for Adobe, bridges the gap between fifteenth- and sixteenth-century calligraphic and eighteenth-century Modern styles. Gotham, inspired by New York street signs, was designed by Tobias Frere-Jones for Hoefler & Co.